Joseph Oler and Jake Blue

ARE WE WORTHY?

Two Brothers' Quest for Adventure, Fortune, and Perspective

AUSTIN MACAULEY PUBLISHERS™

LONDON • CAMBRIDGE • NEW YORK • SHARJAH

Ordering Information:
Quantity sales: special discounts are available on quantity purchases by corporations, associations, and others. For details, contact the publisher at the address below.

Publisher's Cataloging-in-Publication data
Oler, Joseph and Blue, Jake
Are We Worthy?

ISBN 9781645754589 (Paperback)
ISBN 9781645754596 (Hardback)
ISBN 9781645754602 (ePub e-book)

Library of Congress Control Number: 2020909833

www.austinmacauley.com/us

First Published (2020)
Austin Macauley Publishers LLC
40 Wall Street, 28th Floor
New York, NY 10005
USA

mail-usa@austinmacauley.com
+1 (646) 5125767

Table of Contents

This Book

In the summer of 2018, two university boys bought a one-way plane ticket from Spokane, Washington to Charleston, South Carolina with $1300, a couple books, and a few goals. We wanted to see if it was still possible to hitchhike across the country in the technological age. We were looking for Forrest Fenn's treasure, which was supposedly buried somewhere in the Rocky Mountains, and we wanted to suffer in a way we had never suffered before. In this book, there will be profanity, unsightly images/ideals, and a lot of life changing moments from the both of us. I would never recommend anyone try what we did because we are forever changed from that experience. We thought the physical aspect would be the hardest, but the real challenge was beyond anything we could have ever foreseen. Though seemingly different accounts of the same trip, these are two brutally honest perspectives of the country we love, and people within.

Backpack List

Frying pan (cast)   (acutally?) Hats

Camp-stove / Propane bottles x2 (Sterno Cans?)

— Solar phone chargers x 2

2-person tent

— sleeping bag

rain slicker

2-3 shirts

1-2 pants

1 swim wear

1-2 shorts

towel

Shovel

Bolt-cutters ?

— Flint/steel & lighters

— Hammocks

1-2 pair of boots / pair of shoes

3 underwear

5 pair of socks

2 sweatshirts

Cotton underlayer / long johns

The books

New personal books

Journals

knives now

Compass

2 large canteens

— misc. smaller bottles

Safety rope

Fork/knife utensils

Sunscreen

2-3 long-sleeve shirts

2-3 short-sleeve shirts

burner phone

burner wallet

toothbrush

tooth-paste

q-tips

First aid kit

yea? everything goes

Unopened

Fake ID's

Camera

Map

Soap

contacts / glasses

The Dynamic
Between Us

It was on an Alabama bridge in the Southern sweltering heat that Joey told me something that will always stick with me. He said, "Jake, you know why we're friends? It's because the optimists and realists need each other. The optimists are the kite in the clouds and the realists are the man on the ground holding the string. Without the kite, the man would never look up at the sky and see what was possible. Without the man, the kite would stray too far into the clouds to never be seen again." Joey saw himself as the realist, i.e. the man on the ground. I, on the other hand, was the kite in the clouds, the source of his inspiration. He couldn't have been more right looking at the relationship we had before this adventure.

Jake

Joey

About Forrest Fenn and the Treasure

Forrest Fenn grew up just West of Yellowstone National Park. He has written multiple books detailing his adventures as a young man. Those stories include lassoing a buffalo and having it tow his car across the plains, getting lost in a national forest with his friend, and even the adventure of making millions of dollars as an art dealer. The older Forrest got, the more he would see the generations that came after him turn more and more to technology instead of to the vast outdoors that he did as a child. This troubled him, and he decided that he had to do something about it.

Forrest purchased a large bronze chest and filled it with personal treasures he had gathered over the years. According to him, the chest included trinkets like Native American pearl necklaces, Spanish gold coins, and much more. The total worth ends up being somewhere between two and three million dollars. At the ripe age of 79 years old, Forest set off into the Rocky Mountains with his chest in tow and hid it for a lucky someone to find some day. The only clue he left behind is a poem that goes:

As I have gone alone in there
And with my treasures bold,
I can keep my secret where,
And hint of riches new and old.

Begin it where warm waters halt
And take it in the canyon down,
Not far, but too far to walk.
Put in below the home of Brown.

From there, it's no place for the meek,
The end is ever drawing nigh;
There'll be no paddle up your creek,
Just heavy loads and water high.

If you've been wise and found the blaze,
Look quickly down, your quest to cease,
But tarry scant with marvel gaze,
Just take the chest and go in peace.

So, why is it that I must go
And leave my trove for all to seek?
The answers I already know,
I've done it tired, and now I'm weak.

So, hear me all and listen good,
Your effort will be worth the cold.
If you are brave and in the wood
I give you title to the gold

Forrest's story spoke to Joey Oler years before we set out across the country. Joey has a detective-like mindset and a hunger for adventure, like me. Sitting on the bank of the three rivers convergence in Southern Idaho in front of a tiny campfire, Joey introduced me to Forrest's story. I was fascinated that someone would lend such a service to the community just for the sake of adventure. Joey also shared that he had already done some preliminary detective work on where he thought the treasure was located. When I heard what he had to say, I simply couldn't say no to his request to come and help him look for the treasure. I told him, "I'll come with you on one condition, Joey. We need to hitchhike from the East Coast to the Rocky Mountains to find it." In that moment, our brotherhood evolved into something even greater: a partnership.

Jake's Heart
Opening Up About Who We Are

Joey Oler. I didn't think much of it the first day I matched that name with his face. It was sophomore year of high school, a year filled with hormones, social classes of 'cool' and 'uncool,' and the first real thoughts of the future. I had just switched schools for basketball, but I had no idea how out of place I would feel in a jumbled mess of 1100 kids after stepping out of the family of 96 my old school had bolstered. Searching for some form of normality, I signed up for band class. I had taken it at my old school for fun but at this new school, it was going to be my one way to remember what normal was. That was the day I would first meet the boy that would grow to become my brother.

A crash of a cymbal during a lecture caused the whole class to jump and turn to see where it had come from. There, in the back of the room, stood a tall kid with a sheepish grin on his face; a few seconds of silence and that grin turned into laughter that the whole class would soon follow. Even the teacher had to suppress a grin as he walked Joey through the process of putting his drumsticks down. "Yes, now put both drumsticks in one hand; good Joey. Now, lower your hand to the floor and open your hand, allowing the drumsticks to fall to the floor." This was soon followed by applause from the rest of the class as Joey jokingly struggled to follow the simple instructions. I learned then that I was lucky to be in a class that had a great sense of humor, and that this Joey kid might be all right. Of course, I could have never predicted the events that would soon lead us on the greatest adventure of our lives up to that point.

Time went by and Joey and I soon became friends. The thing drawing us together seemed to be that we both exuded an aura of something different that didn't seem to mix with the other kids in our classes. You see, we were doers unlike so much of our generation today. Walking down the halls of our high school and most other high schools, you will see troves of phones pulled out the moment the bell rings. Kids wall themselves off from social interaction to go home and drown in more screen time, more sitting, and a lot less learning how to survive in the real world. Joey and I had reached a consensus that our generation was, for lack of a better word, wrong. All we knew at this point was that we could not become the slaves of a system that we had not consented to.

Senior year rolled around and we got an opportunity to express this pent-up belief. Our band teacher spent his summers in central Idaho as a river guide. During our last year in band, he extended an invitation to our senior class to join him on the Lochsa River. The whole class hmmm and hawed for a couple of minutes, but my fellow adventurer and I just grinned. I didn't even have to look at him to know we were going river rafting that summer.

A couple of months later, Joey and I, along with a couple other adventurous souls, piled into Joey's jeep and began the long drive down to the wilds of the Lochsa River. Conversation filled the car, but before long, Joey and I were the only ones left talking. You see, the differences that had separated us from the normal crowd had also brought Joey and I closer as friends. Those same differences had taken their toll in other areas of our lives. I had switched schools for the opportunity to become a division one basketball player. Thinking that parkour would help me on the basketball court, I leaped over a bike rack and tore ligaments in my left ankle. Two months later and properly recovered; I came back for one day before I tore ligaments in my right ankle. My dream had officially ended, but I was not alone. Joey's first love had just broken up with him for being 'too nice' after he showed up to her birthday in a tuxedo with a promise ring.

We were two boys trying to be adults, but were scarred by the woes of high school. With nothing but each other to encourage us towards our future, we talked. We talked as old men do about war, with stories of our friends' funerals who hadn't been lucky enough to make it to senior year with us. Tales of our families that, despite all their love, just couldn't understand what it felt like to have such big dreams with no idea of how to achieve them. We even progressed into telling legends of ourselves long in the future about what people would tell their children about Jake and Joey. Our friends in the backseat must have been completely engrossed in our talk or asleep, because I don't remember hearing a word for the two hours that Joey and I shared our stories.

What seemed like a millennia later, we arrived at the three-river convergence campground. (Not the real name, but what I will always remember it as.) After meeting our band teacher turned river guide and the rest of the employees at the river rafting company, we settled in for an early night in preparation for the river rafting early tomorrow. With the excitement of sixth grade girls (well, me at least), we jumped in our sleeping bags with dreams of white water ahead.

Waking up early, we rushed into one of the most fun trips I've had to this day. A couple of crashes and a considerable amount of time being soaking wet on the river, we made it back to camp tired and hungry. After scavenging what food we could, Joey and I realized we still had another six hours before sunset. Without a second thought, we wrangled up whoever else that had the energy to come with us and took off towards the mountains.

Flying down dirt roads with country music blaring, we made our way far enough out that there was no cell phone service and certainly no one that knew we were so far back in the woods. Being the geniuses we were, the group decided that it was time for some nature parkour. It started innocently, jump off that rock, slide down that hill, but testosterone soon reared its ugly head and pushed us to more intense feats. Despite our better judgment, we ended up laying a log across some savage looking rapids on top of a waterfall. We all stood at the base of the log for a minute knowing it was a stupid idea, however, logic and teenage testosterone are not friends. 'I bet you can't make it to the rocks!' Followed by an 'Oh, yeah? Watch this!' and we were tempting Fate to end us in no time. (So sons everywhere are not chained in the basement to protect them from their own stupidity, I advise all mothers to skip ahead three paragraphs.)

Like ants making trips back and forth from the colony, one by one we ventured out across those rushing rapids on our sorry excuse for a bridge. The roaring waterfall on our right assuring us that any misplaced footstep would end poorly. Still back and forth we went and not a single casualty was earned. With a smirk on his face, one of our friends would later tell us that he proposed we had enough fun for the day and it was time to go home. Everyone agreed and marched back to the car, but upon their arrival, they found that Joey and I were nowhere to be found.

No, we had not fallen off a waterfall, but it's true that we were nowhere near the car by that point. We had gotten bored with the log crossing ages ago and were already halfway up the maze of canyons and crevices the rest of the mountain had to offer. Leaping, hurdling, and racing, Joey and I bounded up the mountain, leaving the car and our friends further and further behind. We had no idea where we were going, only that we were not going to let a day that had been so fun end without one last adventure. Looking back on all the dumb stuff we did, there is only one conclusion I can come to, Fate must be a mother. The whole crew of kids we dragged out into the woods all could have died walking across that log. Any logical person would take that back home as a story to tell their grandchildren, but we didn't think that quite enough risks had been taken yet. Given that line of thought, there are two ways we can look at what Joey and I tried next.

Example A: 'Wow, natural selection did us all a disservice by letting you guys walk back off that mountain.'

Example B: 'Wow, you guys are going to be the most badass grandparents ever!'

Up the mountain where Joey and I now found ourselves was a geologic marvel. The last Ice Age had strewn massive boulders across the slope when the glaciers carved their way through the mountains. This incredible natural event had left us an adventure wonderland filled with 100ft drops left and right, caves that went down too far to see the bottom, and bizarre animals like pikas that made this their home. The perspective of danger fueled us and we soon found ourselves leaping over those 100ft drops like we were playing hopscotch on the playground and climbing the cliff walls like Spiderman. Like all good things in this world though, it had to come to an end. The danger had not yet been enough to make us turn back, but Mother Nature had a surprise in store for us.

Joey and I had found ourselves at the bottom of a cliff we couldn't climb and were starting the long walk around. Jumping over a pile of shale, we heard the sound of every hiker's worst nightmare, 'shhhhhhh-ch-ch-ch-ch-ch-ch-ch-ch-ch-ch-chhhhhhh.' We had just pissed off a rattlesnake by stepping over its resting place for the day. I can only compare our reaction to those scared cat videos on YouTube we both leaped straight up into the air with verticals that would have put NBA players to shame, and froze.

I vaguely remember one of us saying, 'Don't... move' as we analyzed our options. The ground was mostly loose rock, so any movement from us could provoke the snake into striking if we were not careful. The nearest hospital was over 100 miles away and we had no cell service to let anyone know if we were to be bitten. A rattlesnake's venom is a neurotoxin that slowly shuts your body down, killing you by suffocation. If Joey or I were to be bitten, we would still have to hike down a mile or more, which would result in an elevated heart rate that would suck the venom through our bodies like a vacuum, essentially killing us even faster.

There is wonder in being in a truly dangerous situation that will be covered extensively through our story. The society we have created exists predominantly in a state of past and future, with little consideration of the present. As a consequence, we are raised to think about the past and future while only partially living in the present. Most of us walk around thinking about our car payments, our business meetings later that day, projects that need our attention etc. We rarely see the flowers on the side of the sidewalk, or notice the veins of the leaves that fall in autumn. When one finds themselves in true danger, however, there is a snap back to reality. Instead of focusing on the has and what will be, a person in danger finds themselves fully, and completely, present.

Eventually, Joey and I came to the conclusion that we had but one option, step back over the rattlesnake at our ankles and go back down the mountain. The thought of more rattlesnakes had finally brought us back to our senses. With complete precision, Joey and I leaped back over the rattlesnake. Rattling furiously, the snake's beady eyes watched our every move until we were well out of striking distance. Our state of focus slowly waning, a wave of laughter overcame both of us. Laughing at our own prowess, the fact that we would have died had we played that any differently, and the thought of our friends that were probably pissed not knowing where we were drove us to lay on the ground for one of the best laughs I think I've had in my entire life. It's the little things that make us happy to be alive and that rattlesnake had given us a little thing to smile about that day.

Giving up our adventure, we decided we had better head back to the Jeep and start working on getting home. The hike down was even more dangerous than the hike up, but Joey and I had survived the worst and we were not going to be scared by anything else today. Embodying the spirits of mountain goats, we ran down the mountain. Leaping from rock to rock with skill that neither of us knew we possessed. Coming to one particular ledge that was too big for me to comfortably jump off, I started the climb down. Just as I reached the bottom, I looked up to see Joey switch from the spirit of a mountain goat to one of an eagle and throw himself off the edge of the cliff, well above my head. He landed on a lower boulder hard enough for me to feel the vibration of the impact. He slid and rolled for another couple of feet to narrowly avoid falling off another equally as large ledge and then came to stillness on the ground. My first thought was, *Oh, he dead,* but a moment later, Joey re-embraced the mountain goat and hopped up with a smile on his face. *God, I love this dude!* I thought to myself.

A couple hours later, the whole crew of us sat around the fire on the riverside where we were camped. Campfire stories galore were being swapped back and forth about what we had done and what it had inspired us to do. It came to be Joey's turn to share and he ambitiously threw out the idea of searching for Forrest Fenn's treasure. The other guys all laughed him off while I sat in silence and listened. They said, "There's no way, man, no one's found that thing." Little did I know, two years later, Joey and I were going to be starving on the side of the road with our thumbs out looking for treasure. Somehow, the bastard talked me into it.

Perspective
Learning How the South Works

Trial Run

Summer 2018 was our agreed upon time. That left us two years to plan and condition ourselves to be able to walk 20 miles a day with 40+ lbs. on our backs. Lots of preparation equals being ready for that moment, right? Well, I couldn't tell you actually, we bought the tickets a year prior and forgot all about it until about 12 days before departure... Not the best start to a trip that we guessed would last two or more months. We thought 12 days was still plenty of time to get in shape, pack, do a trial run, say goodbye to our friends, plan our route, call our potential friends scattered across the country for a place to stay, and organize the trip with our banks so we could buy food in every state. Easy, right? Wrong! To be able to pack everything you want or need in a backpack for the next two months is not easy. On top of that, Joey and I had not conditioned AT ALL for the kind of exertion we were about to go through. We determined a trial run was necessary to see what we were up against. It was simple in theory, we were going to attempt a 32-mile walk from our hometown Sandpoint, Idaho to Hope, Idaho and back.

Rules for the trial run:

No accepting rides. We couldn't be sure anyone would pick us up on our trip.

We had to carry our bags. Mine was right around 40 lbs. but Joey's was over 50! Someone had to carry the bolt cutters for those hard to reach places...

Lastly, minimal food intake to simulate not being picked up for multiple days while low on food.

I showed up at Joey's house just before 7 a.m. pumped with excitement. His parents laughed as we walked out the door, but it only added to our determination. Joey and I had both been told we were crazy and that we would never do this for two years leading up to the last week before we left. Any laughing or making fun of us only served as fuel to keep going now.

Reality set in about three miles into our trial run. The difficulty of what we had set out to do became very real. We already had blisters on both feet that made it hard to walk. It got bad enough by mile six that we had to start stopping every mile to pop new blisters so we could walk without limping. Not halfway through our TRIAL run, Joey and I were hurting bad. To force ourselves to finish without succumbing to hitching a ride, we took what we thought would be a shortcut along a stretch of railroad that went across part of the lake. It was very pretty, but it was hard to take in the sights with the uneven ground torturing our blistering feet. Our trial had started much worse than Joey and I expected. Neither of us said anything for the sake of pride, but we were both ready to quit at about nine miles in. I didn't know what kept us going at the time, but looking back, it is crystal clear.

The Fear

We had stopped at a swamp alongside the railroad tracks. Exhausted and hungry, we plopped down to snack on a few granola bars and question our sanity. We were crazy for doing this, weren't we? Nobody else does this, why should we make ourselves suffer? The reality is that we were justifying our fear with logic.

In the world we live in today, fear is a staple of society. From our parents not allowing us to walk to school in fear of kidnapping to the mass media marketing the next apocalypse like they are handing out hotcakes. Fear is everywhere, but what is it exactly? Merriam Webster dictionary defines fear as: 'An unpleasant, often strong, emotion caused by anticipation or awareness of danger.' To paraphrase, fear is the creation of a future moment in time in which our wellbeing will be harmed. By using this logic, we can determine that fear does not exist in the present, only by creating an outcome of the future in our minds can fear have any sway over us.

I do not wish to discount the importance of fear, it is a natural and necessary response for the survival of the human race. However, we now live in a world that is abnormally safe where fear has little basis in reality. Fear evolved to protect us from future danger; take a moment and look at the state of fear in the world today. Fear is no longer protecting us, but holding us back from a myriad of opportunities. Evolution created a need for fear within our ancestors when they were hunted by ancient predators like Smilodon and the Short-Faced Bear. Now, it 'protects us' from walking out our front door because our neighbors' dog might bark at us. Don't believe me? Look at what some of the most common phobias in the modern age are.

- Trypophobia: the fear of holes. (This includes honeycomb holes)
- Aerophobia: the fear of flying.
- Mysophobia: the fear of germs.
- Claustrophobia: the fear of small places.
- Astraphobia: the fear of thunder and lightning.
- Cynophobia: the fear of dogs.
- Agoraphobia: the fear of crowded or open spaces.
- Acrophobia: the fear of heights.
- Ophidiophobia: the fear of snakes.
- Arachnophobia: the fear of spiders.

Just the idea that we have created a scientific word to better define one's particular fear shows that we live in a world filled with fear. Not just instinctual fear that keeps us alive, but manufactured fears that have almost no correlation with danger. Looking at that list above, answer honestly, how many of the people that have these fears run into life threatening danger poised by one of these phobias?

Using our logic from earlier, these moments in which fear exists do not exist in the present! From the moment we are born, we are taught to live in nonexistent times and places with almost no regard for the present. This means that we have an entire population of people that are basing their entire lives on the falsehood that their well-being will be harmed sometime in the near future. It keeps us from being the eagles we were born as and confines us to the life of a duck, never living up to our true potential.

Climbing the First Wall

Sitting next to the swamp on the side of the railroad tracks, Joey and I got our first taste of this real fear. We felt the urge to quit because of the inevitable danger we were going to run into, not on our trial run, but on the vast trek across the country. It took a long while for us to get to this moment, two years of people saying we couldn't do it, two years of people telling us every bad thing that could happen, two years of living amongst the status quo as a duck. In that life changing moment, our differences that had brought us together over four years ago rejected the societal influence that had held us back for so long. We stood as brothers and replaced that feeling of dread with one of fascination and excitement about what we were about to do.

Determined to make our way to Hope, we picked ourselves up off the ground. Six and a half hours of blistered feet and growling stomachs from starting our trial, we trudged into my Hope house's driveway. We were greeted with hysterical laughter from my father who saw how beat we looked after a mere half day's trek. Too tired to say much other than, 'Hey,' we bypassed him to the grail of survival, otherwise known as the kitchen sink, and drank like camels preparing to not see water for the next couple weeks. We had made our first checkpoint, despite it only being half of what we had wanted to accomplish, we had started the day with a success. Now, we only had 16 miles to go...

A couple of rest filled hours later and we were off back to Sandpoint, but our aching legs and blistered feet had other ideas. We walked for two miles before we hit the highway and put our thumbs out, breaking the first rule of our trial run. We thought we had worked hard enough for the day and needed to give our bodies a break. The mental aspect, however, ended up hurting us a lot more when we flew across the country. A couple of minutes after putting our thumbs out, a kind man stopped and gave us a ride straight back to the doorstep of where we had started. Growing up in rural Idaho, one could expect this behavior. We actually had to turn down rides on the first leg of our trial run. This set a subconscious thought in my mind that whenever we thought our trip was too hard, we could be bailed out by a friendly driver. I would soon come to find that was not the case where we were going.

The Plane Ride

A couple of days later, Joey and I woke up at 3 a.m. to meet at the airport. We hugged our families goodbye and watched them drive off into the still dark early morning. There were no tears shed that morning at the airport, Joey and I aren't usually much for crying, but we were ready to fend off the tears of our much-loved families. The fact that they didn't cry gave us the ultimate chip on our shoulder, because at that time, it showed us that no one, not even our own families believed we could do it. There was no turning back now. Joey and I stood shoulder to shoulder as brothers who would live and die together through whatever was to come.

Walking through the airport felt like walking on air, nothing seemed quite as real as it should be. We both felt the adrenaline start pumping when we handed the South West lady our bags. Despite it being 5 a.m., we were wired and ready to go. On that day, I started a quest of back flipping in every airport I set foot in, I'm now at six as of writing this.

Walking through security, we started to notice the differences between all the other travelers and us. We watched them hand over huge suitcases to the attendants, some were dressed in suits. You could tell they were business people headed on some trip for a weekend or two to impress somebody higher on the social hierarchy than them. Then there were the vacationers, you could tell them by being dressed for comfort, some even still in their pajamas to sleep on the flight. You could even tell what kind of trip they were going on by the little mannerisms like the way they walked. The business men scurried like mice, always in a hurry no matter how early they were. The travelers would walk in morning zombie mode; they were exhausted, but there was an air of relaxedness about them in their sleepy smiles and dragging feet.

I wonder how we looked in the sea of people in the airport that morning. Everyone else seemed to have their entire lives in their multiple bags and suitcases. Comparatively, Joey and I were garbed in a button up hiking shirt, a pair of breathable pants, a sun hat, and some small hiking backpacks compared to the huge suitcases the rest of the travelers had. We looked completely estranged from society and doing backflips down the halls didn't exactly help our image. It was still early morning luckily, so we only caught a few sideways glances from security guards on our way to the plane. We made our way upstairs and watched through the airport windows for our plane to come in. We didn't have to wait long to see that blue and red plane pull up to the gate. I had a tingling feeling that spanned the length of my spine as we were ushered down the hall to our waiting seats, the vibe finally changed from one of goofy excitement to accomplishment. We hadn't done anything yet except step on a plane. For us, though, this was saying goodbye to the boys we were and beginning the journey to become the men we wanted to be.

We Are Here

Flying down into Charleston, I felt the butterflies build in my gut again. As the wheels touched down on the runway, I knew we had started something that we couldn't quit. We were now stuck in South Carolina whether we liked it or not. For those of you who have been to both sides of the US, you will know they are vastly different. Joey and I had heard stories of the different environment and the different kinds of people we were going to meet, but we soon realized that people can tell you stories all day. Until you experience it, you have no idea what you are getting into.

Stepping off the plane, we had one goal. Somewhere in this massive city was a hotel for our first night. It was about 4 p.m. which would give us plenty of time to wander around and find it. We found our way down to baggage claim and grabbed our bags. Flipping them over our shoulders, Joey looked at me and said, "You ready, brother?"

"You kidding me? I was born for this, Jo Jo." With a grin, we walked through the front doors of the airport to our future.

"O…god," we said in unison.

As soon as the doors opened, we were blasted with a wave of humidity that took us by complete surprise. Growing up in North Idaho, we had never experienced the feeling of swimming through the air. We were still okay, we just had to find our hotel and figure out where the hell we actually were. I swear, we were less than 50 steps in before we were absolutely lost again.

Me: Google said the hotel was west of the airport, right?

Joey: Yeah, I think we're headed north though. Wait… South maybe?

We had grown up with mountains as our compass, now we had a swampy jungle with no elevation to tell us which direction we were headed. Being lost in front of the airport was not how we thought this trip was going to start. We tried to ask someone where we were, but like us, they were all travelers with no idea where they were. The lost feeling got old pretty fast and we resigned to the fact that we were going to need a taxi to figure out this mess of a city.

The Locals

We flagged down a nice man that was very entertained by what we were wearing. First thing that came out of his mouth was, "Y'all ain't from around here, are ya?" The southern accent was heavy. Joey and I looked at each other again, "Dude, siiick."

Turning back to the nice man, I said, "No, sir, we are not and we're already lost." That was all he needed to hear, and we were flying down the grid of roads that is Charleston a second later.

Our first ride did not go as expected, but it was educational. Apparently, hitchhiking is a novel concept in South Carolina. For when the man asked us, "Why the hell yous boys dressed like Indiana Joneses?" I answered, "Well sir, my friend Joey and I are from Idaho. We bought a one-way plane ticket here and are trying to hitchhike back."

Driver: *crickets* I'm sorry, y'all trying to hitchhike back? Like thumbing rides?

Me: Yes, sir.

The most awkward silence I have ever had then commenced. I waited for him to say something for a minute…*silence*…*crickets*…*more silence.* I looked to Joey for help.

Joey: So, uh, you ever been to Idaho before?

The man looked at us out of the corner of his eye and took a couple more seconds to let the soul crushing silence envelope the car.

Driver: Nah, man, I've been close to y'all though. I got friends in Michigan 'n stuff.

Joey and I made eye contact for what felt like the 100th time that car ride. We whispered, "He means Iowa, right?" We nodded in approval and looked forward again. After a couple seconds of silence, I opened my mouth again.

Me: No, sir, we're from Idaho not Iowa.

He looked at me outta the corner of his eye again.

Driver: That's what I said, Iowa.

Joey was on the brink of busting out laughing, but he caught my eyes pleading for help again and quickly composed himself.

Joey: No, sir, we're from Idaho. Up by Washington State.

Driver: *Soul crushing stare* Y'all saying yer from that potato place up there?

Not sure of what my response should be and why he was looking at me instead of the road, I decided that nodding my head would be the safest response. This was met with hysterical laughter by our driver. The rest of the ride passed in silence.

Getting to our hotel $20 poorer, we told the check-in lady that we had a room reservation from three weeks ago. Her curious eyes looked us up and down. "Uh huh," she mumbled, then turned back to her computer screen. _We assumed that meant she was getting our keys, so we waited. Just before she handed us our keys, a maid walked into the front desk area that was not as subtle with her shock of how we looked.

"Y'all ain't from around here, are ya?" We then proceeded to have almost the exact conversation with the maid as our taxi driver.

"Uh, no, ma'am. We're from Idaho and trying to hitchhike back." *crickets*

"Y'all mean Iowa?" Etc. etc.

In our hotel room later that night, Joey and I were pouring over at every map we could, trying to figure out how the hell we were going to get out of Charleston. Google maps said we had easily another 15 miles before we were out of the beehive of people. This could have been a very depressing moment for a lot of people, but our excitement came back to us at the perfect moment. Something about the feeling of starting an adventure, the sense of freedom and breaking away is irreplaceable. It is my favorite feeling in the world and Joey felt it too, we went to bed fully excited to suffer like we had never suffered before.

Starting Out

The sunlight woke us from the last blissful sleep Joey and I would have in a long time. I remember smiling like a fool waking up; chills of fear and anticipation were running up and down my spine because today was the first day of our long trek back home. We helped ourselves to the most disgusting hotel breakfast I had ever had, literally powdered eggs that you could still feel the powder in, and walked out the doors with a smile on our face.

We stepped out those doors into a new and potentially hostile environment. We were lucky, however, Joey and I had three different stops planned in as breaks from the road (could have had more, but our lack of planning had snuffed that possibility). I had two cousins in the South that would take us in, and Joey had a long-lost aunt that had also said she would love to have us. Stop #1, my cousins in Lexington, about 120 miles away from our hotel.

With no mountains to guide us inland, we were restrained to Google maps to find our way around. After we found a heading, I remember vividly the first step of our hitchhiking adventure. We were standing on the side of a four-way street fully garbed in our hitchhiking gear. Each of us in mostly tan clothes and an Indiana Jones hat. On our backs were what was to be our only lifeline for the rest of the trip; our backpacks. Our backpacks represented the most planning we had done up until getting on the plane. The stuff we piled in those things was crazy. Each of us had an extra shirt and an extra pair of pants, two underwear, two pairs of socks, a sleeping bag, a water purifier, $1300 on a debit card, our phone, a knife, three water containers, and some medical supplies split between us. We had also coordinated our gear so that we both had things the other needed. I carried our tent, a solar phone charger, and a book of the railroad routes in the South (don't worry, I'll get to that nightmare of a story later). Joey had our cooking ware, a portable stove, and bolt cutters in case we wanted to get in some real trouble.

All of this was running through my head until Joey snapped me out of my daze, "You ready, brother?" I swallowed and took one last look around that street corner in South Carolina, the sun was shining through the leaves, the potholes in the driveway leading to our hotel, the way too perfectly cut grass. All so I could remember that day as when my new life started, or as the day I condemned myself to the very real possibility of never coming back.

Acceptance washed over me of what was to come, I looked back to my best friend in the world, "I'm always ready, Joey." With that, we took our first step together into the great unknown.

Information Gathering

The year before our hitchhiking adventure, I was attending the University of Montana in Missoula. Like most college towns, it is a cultural mixing pot of people from all over the world. Going to that university gave me the opportunity to socialize with all different kinds of people. The year before I went on the road to hitchhike, I realized I could use these people to social network and start learning what sort of environment I was going to find myself in on the East Coast. In talking to these people, I found two main points between all of them. One, the landscape is nothing like the West. They told me about ticks that carry Lyme disease and blistering heat from the humidity. Two, I was told over and over that the people would not be the same as they were in the West. Of course, my first question was always, 'How will they be different?' What they would tell me was that, 'People are a little more in their own worlds over there.' I didn't really understand exactly what they meant by that, so I tended to laugh that crucial bit of information off. A small-town Idaho boy could have never predicted the gravity of what tip #2 actually ended up meaning.

Questioning

Joey and I had started an amazing journey with those first steps, but the difficulty was far beyond what even our trial run prepared us for. We started off our walk with enthusiasm, I cartwheeled into a moonwalk across our first street light, expecting some smiles from the people in the cars, I turned around to give them a little bow. Instead, I made eye contact with something I hadn't ever really experienced until that point. The people looking back at me didn't have the glimmer of amusement in their eyes like I usually got when I was up to my antics, their eyes spoke of fear and skepticism. Out of the corner of my eye, I saw a lady take a picture of us, thinking I had found a good audience, I swung my attention her way and smiled. When she noticed me looking, she responded by shoving her phone back in her purse and scurrying in the other direction. Without much thought, I labeled them as a tough crowd and jogged to catch up with Joey. I hadn't put all the pieces together yet, but that was a very small taste of tip #2 from my friends back in Montana.

Joey and I continued on our merry way, making jokes and doing our best to laugh off the not yet healed blisters on our feet from our trial run. Three miles into our journey, we ran into our first opportunity to add some spice to our trip. Walking over a bridge, we looked down to see that this was our first railroad track crossing! Not only that, but a train was stopped right under us getting ready to take off again. Joey saw it as the opportunity it was and instantly looked at me with that devil's spark in his eye, "Should we?" he asked. I returned his smile and got my hopes up for a moment thinking, *Tomorrow, we could end up anywhere from Maine to Kansas if this train takes off with us in it!* I dowsed that spark of thought almost as soon as it lit. There was a part of that thought that was exhilarating. Being on the open railroad as it ferried us to who knows where sounded incredible. There was a feeling I couldn't shake that kept me from following Joey down to those tracks though, part of me felt like I was cheating. Not five miles into our journey and we're already looking for ways out of it? I don't know what it was, for some reason, I felt like we needed to really suffer before we earned any sort of reward like the railroad. I convinced Joey to come of the same thinking and we continued on our walk. Let me tell you, my wish to suffer a little more was on its way.

Hours started ticking by and the sun rose to its reputation in the late morning sky. The 85% humidity in the air combined with the 70°F heat made for a killer on the legs and back. I had decided that the rubbing of my shoes on my healing feet was the problem, and if I just went barefoot, my feet would feel better. Joey warned me against it in the city littered with broken glass along the side of every street. Of course, that wasn't going to stop me. My reasoning? "I've cut my feet open with glass before, Joey." Only looking back now do I see how stupid I sounded saying that. Not one mile later, I felt the smooth incision of a blade of glass slip in and out of my foot. I suppressed the cry that came to my tongue as to not let Joey know he had been right. Plus, we still had an estimated 2,794 miles to go (actually ended up being a couple hundred miles further) and a little cut was not going to stop me. I convinced Joey to stop at a bus stop bench so I could quietly slip my shoes back on. I snuck a quick look at my cut, it wasn't bad and only bleeding a little. I could tough through it, but it wasn't going to be easy. Day one on the road and already my ego had cost me being able to walk properly.

I let Joey take the lead so he wouldn't see the limp I was trying to hide. Luckily, we were both out of water and getting hungry, that warranted another stop at the next grocery store. Everything was different in the South though. We stopped at something called the Food Lion that looked like a grocery store, but you best believe Joey and I were not about to walk into a lion den on day one. We spotted a family walking over the front doors and waited to see if they would be devoured by lions. The doors closed behind them and we held our breath waiting to hear the screams… "Nope, must be good!" I exclaimed. Despite our aching feet, Joey and I were always able to laugh at nothing together.

A Study in Pink

We resupplied and were just getting ready to begin on our way again until the cutest little old lady walked up to us. She was dressed in all pink from her golf cap down to the pink laces on her shoes. The kind of person you see across the parking lot and get the instant urge to run towards her, vault a car, slide under a moving semi, and give her the best hug of your life. I'll always remember her as the first generous soul we met on our trip.

With a smile, she grinned at us with the kind of smile that shows the years of happiness she's been through. "What on earth are you two boys laughing about looking like this?" I don't remember what we were laughing about, but that lady gave us about 10 more reasons to keep laughing. Aside from the pink and her extreme southern accent, she was one of those stubborn old people that uses all those funny old metaphors.

She asked us what we were doing and of course we answered, "Well, ma'am, we're hitchhiking across the country."

She laughed, "Well, steal my dentures, string them up over an alligator pond, and charge a $10 fee for the wrastling show. Y'all are crazy!" Any attempts to not laugh at this lady were quickly thrown out the window after that one. We chatted for a couple more minutes before we told her we still had a long way to go before we got home. With a chuckle and the same youthful smile, she shook our hands and let us be on our way.

It's funny how when you go to take a rest, you feel like you made a good decision until you try to stand back up. It had been a mere half day since Joey and I started our trip and our bodies were letting us know what they thought of the idea. My foot where I stabbed myself was on fire and Joey had almost 20 pounds more in his bag than me on day one that was killing his back. So much pain in only about five hours didn't even make sense to us. I think part of the surprise at our struggling was because we thought that having six pack abs made us invincible (it doesn't). No matter how bad it got on day one, however, neither Joey or I ever complained about any pain either of us had. When pain got too bad to stay silent, we laughed. When it was too bad to laugh, we talked about how good this trip was for us. When the pain finally got too bad to even talk about the good, we talked about all the people that said we were never going to do it, yet here we were. The thought that we could prove the world wrong, and maybe even part of ourselves, was what kept our thumbs up and our feet moving for the next two hours.

It was coming to the point where Joey and I thought we were going to need another break, but it turns out the good stuff you put out in the world does come back to you. A red van honked at us on the way by, hardly the first time that had happened so far, but as we watched it drive up the road, it turned into a driveway and stopped. The window on the driver's side rolled down and out popped the same pink golf visor we had talked to at the store. "Hurry!" she said, waving her arm. Our legs spurred to life and we ran to our Pink Lady.

I will admit I was nervous to get in her car. I've found there are few 80-year-old ladies that drive much better than elephants climb trees, but she surprised me and totally restored our faith in mankind. Of the thousands of cars that day that had driven past us honking, yelling, and flipping us off, this lady put her total trust in us and gave us a ride. It takes a special person to pick up a hitchhiker, and like she told us, it doesn't have much to do with the hitchhiker themselves, more so life experiences the driver has been through. She told us stories of back when she used to hitchhike with her many men. They would go all over the East with nothing more than a couple bucks in their pocket, almost completely living off the generosity of other people. Not to say that she had nothing to offer. She paid the most valuable currency for the free meals and many beds she got to sleep in, open and honest connections with people. She told us that she had friends all the way up and down the coast from over 50 years back when she used to hitchhike. Incredible, a big difference from now, when it is rare to know all the neighbors on your street.

She had us smiling all over again, "That was exactly what we needed to hear, ma'am, thank you."

She stopped smiling, "O heavens, dear, I would never try what I did again today! My advice to you, get home to your families as fast as you can." There it was again, this woman had given us nothing but love and smiles until that moment. Underlying that love was still fear of the world around her, not as much as we had seen in everybody else, but still there. She dropped us off about 15 miles later and with a hug, we said goodbye, never to see her again.

It's crazy how much the mood or state that you find yourself in affects your reality. Joey and I were both getting ready to tap out for the day before our Pink Lady picked us up. After she dropped us off, we hardly needed to take a break. Our legs most certainly complained as we started walking again, but we were far too happy to let that drag us back down. It was only about three o'clock, that meant we had a whole five hours left at least before we had to set up camp. The Pink Lady had left us a lot to talk about on the next leg of our trip. What was this fear we were experiencing? I understand picking up a hitchhiker is often a scary moment for people, but this was different. It was not fear of us, more of an underlying wariness of other people. Even after we had earned the Pink Lady's trust and she had sided with us, she became scared for us about the other people we would inevitably run into. I didn't understand it yet. Little did I know, our grappling with understanding the fearful mentality was to set the stage for the rest of our trip.

Shay

Neither of us knew what to make of our new environment. Day one on the road had already become the longest day of our life, experiencing just about every emotion from joy to hopelessness. Every person we met was giving off peculiar vibes that were different from the way anyone had ever treated us. We shrugged off the flood of thoughts running through our minds and focused on the task at hand, getting another ride before nightfall.

We had made it well out of the city now and the stream of cars had come to a trickle. The good news was that we weren't getting flipped off as much anymore, the bad news was that there were a lot less people to pick us up. Another hour passed and the hope that we were going to get another ride was quickly fading. The second that we were about to sit down for another break, a blue Ford truck flew by and swerved off the road in front of us. Two rides in one day? Hope surged through us again and we found the energy to run to the blue truck further up the road.

I hopped in the front seat of the truck and Joey hopped in the bed. I was greeted by a man by the name of Shay, another former hitchhiker. The man told us a similar story to our Lady in Pink. He had been all over the East Coast with next to nothing to support himself. Traveling here and there, making friends along the way. Now he had a little farm to grow his own food and make friends as they come through his stretch of the state. It seemed like a peaceful life.

I've found that meeting a stranger can be one of the most wholesome experiences you can have, especially when they don't think they'll ever see you again. Both people that had picked us up opened their lives in a way that takes months or years for the people you meet anywhere else. Shay told us stories about being a young man and going to the corner store to party. Walking in, he would see bags of weed and cocaine littered around the house. He told us about where they would have secret spots to hide their drugs from the sheriff when he would come check on them; under the floorboards, behind the fridge, secret compartments behind pictures on the wall. After getting past the initial shock of the craziness he had been part of, it was easy to see the goodwill behind this man's insane stories.

He could only take us about six miles before his turn to go back home. We slowed down at his driveway, it had been a short but very happy ride. I was just getting ready to shake his hand before he looked at me and smiled, "Why don't I drop you guys off at the old corner store a couple more miles up the road?" I wasn't about to say no to a couple more miles, but Joey and I had made a rule to not to be a hassle to anybody.

I tried to say no but Shay was insistent, "Boy, I know what it is to be a hitchhiker, and as a hitchhiker, you never turn down a ride." Unable to argue with that logic, I rested my case.

This became something that would happen often as our trip went on. People would almost always tell us, 'I can only take you to the next exit.' Or, 'I'm on my way home so I can't take you far.' We had no problem with that, every ride was a gift and we would treat them that way. However, it was common our drivers would get to their destination and keep driving with us. By getting to know people, we were giving them something that we are so deprived of in the age of technology; Human connections. The lessons that were being piled onto us after only one day were becoming too much to process. Joey and I had a goal to hitchhike across the country to see how tough we were. Not more than 10 hours into our first trip and we were realizing that there were so many more possibilities for this adventure. By giving our pure intentions to the battered souls across the countryside, we were awakening a side of them (and ourselves) that was starving for good human contact. No arguing, no anger, just happy people trying to stay that way in a world of bad news.

I could feel the tears well up in my eyes after we reached the old corner store. It was such an honor to be brought to a place that was so close to someone else's heart. There was nothing but a couple of brick walls left standing but I could tell the meaning of this place by looking at Shay, his eyes were glazed over and mouth slightly ajar as he relived the past he had in this spot. With a smile, I reached over to shake the man's hand, "From the bottom of my heart, sir, thank you for everything you shared today." I watched him pull himself out of the past to make eye contact with me. His eyes expressed more than words ever could, he took my hand with his vice-like grip and nodded. In a second, Joey and I were all alone again.

Then Came Stretch

We managed to walk another three or so miles before our good mood dissipated and was replaced by the survival gene yet again. The humidity in this heat was just brutal, but we had to make it to the trees we could see in the distance before we took a break. We trudged through the heat to the trees and were greeted by the nice cut grass of the neighbors' front yard. We were too tired to worry if anyone was home and plopped our bags down, hoping they wouldn't mind. The shade gave us the first relief from the sun in a long time, *How nice it would be if I just closed my eyes... NO!* I thought. *One of us has to stay awake.* On cue, a gentle gentle breeze blew through the air as if to say, 'It's okay, Jake, just close your eyes.' It was no use fighting it, my dreary eyes closed and took me to sleep.

"You boys want some water?" My eyes fluttered open underneath my hat pulled over my face. What had woken me up? "You boys want some water?" the raspy voice repeated. Realizing someone was there, my eyes snapped open and my hat flew off my face to see a tall, old man standing in front of me. He was wearing an old military hat, a light-yellow shirt with an American Flag on his chest, and a long white beard that didn't seem to speak wisdom like so many beards do but instead, it spoke of a survivor.

The stuff I say when I'm woken up has since become an inside joke between Joey and I. I woke with a, "Yo, my man, what cha doing here?" Lol. First off, this was clearly not the kind of person you say, 'Yo, man' to. Tattoos on his arms and yellowing teeth portrayed a, 'I don't take no shit attitude.' Second, if I had looked behind me, I would have seen his house and realized I was laying in his yard... Ooops. Out of the corner of my eye, I glanced at Joey to see if he was still sleeping but he was wide awake, staring with curious eyes at the man who would change our perspective of the South.

Around this time, I looked down to see that the man was carrying a jug of water with him. "I saw you two boys out here and thought y'all might need a drink," he said. A little taken back by this rough exterior of a man offering us such kindness, Joey and I stole a quick look at each other.

We were afraid we looked like beggars already on day one, "O thank you, sir, but we don't mean to be any trouble to you. We can be on our way in just a moment."

He giggled, "I got no problem with y'all, take some water." *Interesting,* I thought. We were suspicious, he had done no harm yet though so we thanked him and fumbled through our bags for our water bottles. He introduced himself as Stretch, a name that one of his old military buddies most likely gave him, and what a story he had to tell.

Stretch was a product of the Vietnam war. A young man living a great life in the post WW2 era one day came home to find a letter waiting for him that would send him across the world to fight yet another war. Unlike World War Two, however, the Vietnam War was not a war of moving lines in the sand to push the enemy back. This was a war of attrition that rewarded the side that could cause the most death and chaos.

Stretch, unlike so many others, survived this brutal reality. He spun all these stories for us while we drank the delicious ice water he brought out and listened intently to every word. He seemed like a good man doing his best to deal with his own demons until something a little different slipped out of his mouth.

Originally, Stretch was from Charleston, but it just wasn't the same for him when he came back. Now, he lived around 60 miles outside city limits with his dog. We were having a good time talking, so naturally Joey asked, "What was so bad about Charleston that made you want to move out here?"

He looked to the sky, "The noise," he said.

"Ah, that makes sense. Growing city, lots of people moving in, I would want to leave too, I responded."

"No, no, no," he jumped in, "it was all those damn blacks with the boom boxes on their shoulders," said Stretch. Joey and I went silent. *What did he just say?* I thought. I looked at Joey and saw him thinking the same thing.

We were from North Idaho, not the land of potatoes as so many people label it, but the land of happy and friendly people. Neither Joey or I had ever experienced much racism. Sure, we had read about it in our school books and heard the occasional joke that wasn't all that nice, but it all seemed like a thing of the past. If anything, I took the 'racist jokes' at school as a prime example of how ridiculous the idea of racism seemed. How could anyone participate in such ideology? I've never understood and because of that, I thought racism was a myth of sorts. Stretch had just given us our first very small taste of the beast that is racism. We nervously laughed our way through the rest of the conversation and took the first opportunity to exit.

Walking away, Joey and I were in a little bit of shock. "Dude, racism is still alive? This is crazy, man." We talked for a while about the topic and chalked it up to a one in a million chance we run into an old man that is still racist. There can't be that many other people down here that are like that, right? I'll give you a hint, we were so wrong. Racism was going to come back to haunt us.

The First Night

The sun was sinking and we needed a place to sleep. We walked until we found a part of the road that was not lined with houses and stole off into the woods just far enough back that a passing car wouldn't notice us. Joey made us a freeze-dried meal for the first night with our solar light to assist him. It was about as nasty as it sounds, but our hunger served as seasoning enough for us and we downed it.

It was finally dark after our 14-hour introduction into our new life for the next who knows how long. Our aching bodies were punishing us for our determination to keep going without mercy. Everything hurt, from our shoulders where our backpack straps had dug in, to our knees that dealt with our added weight on our backs, and of course, our bloody and blistered feet. I have the scars to this day on my feet from that first day, this was going to be so hard if things didn't change soon.

I was just settling in, hoping for sweet dreams of getting rides the next day when Joey shook me back to our reality of the woods in South Carolina. "Jake… look," he whispered with fascination in his voice. My annoyance was outweighed by my curiosity, I unzipped my sleeping bag and stepped out of the tent with Joey. Like most good things in life, it took me a minute to realize what I was lucky enough to see. A flash here, then another out of the corner of my eye, suddenly one right in front of my nose. Fireflies, we were standing in the middle of a swarm of fireflies. I had heard stories and seen the movies, but to have it right in front of my face was so different.

"O my god, Joey… it's beautiful." I didn't have to look at him to feel him smile behind me. I stayed up a few more minutes to watch nature's show, it really was incredible. I couldn't let it keep me up too late though, we both knew we still had many miles to walk in the morning. With smiles and warm hearts, Joey and I settled in for bed.

High Alert

I awoke in the middle of the night with a jolt, something wasn't right. What was it that had woke me? Then the sound of a car door slamming shut snapped all my attention in its direction. A car had pulled into a driveway close to us, way too close. I woke Joey up without a second thought. *Why are they here?* I asked myself. We could hear the non-distinct conversation as they climbed out of their vehicle.

Every fiber of my body came to life as my mind went to the worst-case scenarios. We could be on anyone's property and might not be welcome. More than that, they might do something to show us we're not welcome. Joey was fairly relaxed and maybe even a little entertained when I unsheathed my knife. This was a totally new experience for me. I had spent dozens of nights in the woods growing up, often without a tent. I had heard noises in the woods before, a branch breaking, an owl calling, I knew I had nothing to fear from the outdoors of Idaho. Even the bears and the mountain lions that frequented many places I camped didn't scare me near as much as this moment did.

These were people, not animals, and people have many different possible motivations to be here than the animals. Greed, cruelty, vicious ambition, they were all possibilities for those people so close to our tent. Our earlier encounter with Stretch had not set a pleasant tone for the rest of the day and I was fresh out of trust. In the woods back home, I could count on the animals' pure intention of survival, I had no idea what these people were up to. I even went as far as to unzip the tent and walk out into the dark forest to see if I could get a better view of whoever had disturbed my slumber. Joey had to be annoyed, he knew we had nothing to be afraid of. I was not to be soothed by words, however, only silence would calm me down at this point.

I stalked through the forest, knife in hand until I got a view of the people who were talking. Crouching down low, I analyzed the situation for any danger. It was a man and a woman in front of a big black truck talking to someone on the porch of a nearby house. It looked like they had just gotten back from a date and were maybe talking to the mother on the porch? I wasn't close enough to make out particular words, but I decided that I had nothing to fear from these people.

I could hear the voices still talking when my weariness brought my head back down on my makeshift pillow of piled clothing. I still could not sleep, every artificial noise snapped me back into high alert. I did this for a long enough time to become exhausted, the sounds of people kept me wondering of their intentions. Eventually, I told myself that I still had to keep up with Joey tomorrow, and not being rested would not keep to that promise. I looked back into the forest and found reassurance in the occasional firefly still flashing. My paranoia would have to rest tonight.

Thinking back on that night, I think about how scary it would be for those people to know that there was a barefoot and shirtless hitchhiker watching them with his knife drawn that night. I had absolutely no intention to hurt anyone, my own paranoia was just keeping me awake. Still, I think, have there been people in the woods that have watched me at night?

Flashback

Let's take a trip down memory lane, I feel like it is extremely important to understand the mentality we came into this trip with. Joey and I had both gone through a time of big change by going to college. In going to college, you get the option to purge the side of yourself you didn't like in high school and be remade into whatever you want to be. I had spent my entire high school career as a student of basketball and as a consequence, I knew very little about just about everything else.

In sixth grade, I moved up to the junior high basketball team, not because I was more skilled than the other kids, I was just more athletic than the others my age. I never had to work hard at anything and I was expecting the same treatment at the next level. That expectation was flipped on its head within the first couple days, from me getting rocked by all the other kids in everything I thought I was good at. One play in particular, I got a steal and started dribbling down the left side of the court like a mad man. It would have been my first basket in a game, but because I couldn't dribble with my left hand, I got the ball stolen back just as fast as I had gotten it.

Everyone has one of those 'THIS IS IT' moments when you decide that you have let yourself be not good enough for too long. 'From now on, it will never be like this again,' we say to ourselves. In sixth grade after that play, I had my 'THIS IS IT' moment. I sat out the rest of the game and refused to go back in. I went home that night, picked up a basketball, and didn't put it down for the next 5 and a half years. It was a complete obsession and it became all I knew. I chose it over girlfriends, best friends, and even elected to stay home from family vacations so I could practice basketball. Like I said earlier, senior year it was all taken away with torn ligaments in my left ankle, two months later, my right ankle as well. I had put all my eggs in one basket and dropped the basket. As a result, college was a completely new world, I didn't have an identity to express, so I made it up on the fly. College was the buffet of life and I stuffed my face with a lot of shirtless barefoot sprints across campus and a caring too much attitude that tried to be everyone's friend. Two years later, the personality to set out on this trek across the country was a very happy and childish one that thought if I was happy, the rest of the world would be too. O how wrong I was.

So Much Time

We woke to a Southern sun shining through the morning fog, a truly beautiful sight. It was just before 8 a.m. and we had to get a move on. Doing our best to pack up our stuff neatly, we wandered out of the woods back onto the road. *Maybe early morning hitchhiking will be better?* we thought. That hopeful philosophy was shot out of the sky like a WW2 bomber being targeted by anti-aircraft fire. 'Boom...Boom...Boom, mayday mayday, HQ, we are going down, 'VeeeerrrrrrrrrrraaaaoooooooooooooooooKABOOM!' Yeah, that bad. Our feet did not magically heal overnight, now walking on them was like walking barefoot on the grated boardwalks of boat docks, except Joey and I were doing it with holes in our feet from blisters and glass wounds. 'This is what we wanted,' I kept telling myself, 'we can do this.' I knew the days were still the same length as when we left Idaho, we had not urged the sun to stay up late to let us walk longer, then why did this feel so long?

Life was becoming very present and very real for me and my partner. The lack of stimulation was adding to our agony. We spent hours at a time looking at the road ahead of us with nothing to do but keep walking. We tried talking to pass time, but one can only talk about how pretty the fog was for so long. We were confined to silence before long, with the only noise being the occasional groan or stumble from one from one of us. We had been using the chips on our shoulders as sledge hammers to beat through the bigger and bigger walls we kept hitting, but now that we were all alone with no connection to the outside, it was extremely hard to use other people's doubt for motivation. First, our feet hurt, then the morning dew got our feet wet, then the mosquitos came out to feed. This continuing pain and lack of stimulation put us in a very unique situation. For the first time, we couldn't use the outside world for motivation; the desire to keep going had to come from inside.

Normal day to day living in the 'societal world' (as I took to calling it while on the road) is filled with constant stimulation. Think about what the average American does on a typical day. They are either in school listening to people speak that they would rather not listen to, then hopping on their phones the minute they get out of class to text their friends how boring their last class was, or they are at a job trying to make ends meet, working all day for a purpose they have long forgotten. The stress is so much that the only way to deal with it is to come home and turn on the TV so they can shut their brains off and try to forget the voice telling them that what they are doing is never going to get them anywhere.

The life of, not just Americans, but most of the world is a life filled with chasing stimulation to escape a reality they would rather not be part of. With so much entertainment value at their fingertips, how could they not watch Lele Pons latest video of what it's like to go to the gym with her friends or see LeBron James throw down his 100th dunk of the season? Joey and I were very much living the same exact way before this trip. Essentially just existing and waiting for the future to get better, while we're waiting though, might as well shut off our brains and speed time up a little, right?

Our first day on the road had stripped us of that lifestyle and day two was going to do it's best to break us. We no longer had the stimulation at our fingertips; just us and the road. I can hear someone asking, 'What about all those boring lectures in school though? People turn their brains off all the time for that, why couldn't you do it on the road?' It's a fair point, everyone has their off mode when we are listening or doing something, we don't want to be a part of. It was exactly what Joey and I were doing for the first couple of hours on the road. Day two, however, brought us the X-factor of pain that didn't allow us to tune out reality. Every step was painful, not quite enough to make us stop, but definitely enough to feel it. The heat and the constant exercise made it feel like our bodies were sweating out our fluid faster than our full stomachs of water could process it. It didn't help that every hour, a couple of cars would drift over the divider and speed directly at us to scare us off the road or flip us off. Out of all our days, it is hard to choose which one was the hardest. The misery of day two though? It would definitely rank up there as one of the worst.

Innovation Time

Though it felt like an eternity, it couldn't have been too long walking that morning before we were picked up by yet another former hitchhiker, they seemed to have sympathy for us. He was an uber driver headed to golf with his pals about 20 miles closer to our destination. "THANK YOU!" The guy was fantastic as were most people that picked us up. Saved us a whole day's worth of walking and put us within 50 miles of Lexington, where my family had said they would take us in. Before we hopped out of the car, we asked him, hitchhiker to former hitchhiker, "Any advice to help us get more rides?"

The man smiled, "Get a cardboard sign." Then he was gone. We grinned at each other, *Of course!* we thought. *We will definitely get more rides with a cardboard sign that says where we're going!* Revitalized with a shot of hope, we jogged to the gas station up the road to look for cardboard.

The look on the people manning the register became a staple of our gas station visits all across the country. Only one of us could go in the store at a time while the other watched our bags out front. We would walk in literally dripping with sweat and shoot a grateful smile at the register person to see what they would do. Their faces were priceless, they would usually start with big eyes of surprise that progressed into a moderate suspicion or fear as they kept their eyes glued to us the whole time we were there. It became a bit of a game for me to bet on what each person's face was going to be like.

Joey grabbed a couple of snacks from inside and then it was my turn to find our cardboard. They didn't have any inside, but luckily a Lays delivery truck pulled up to drop off chips at the gas station. I ran over to them as they pulled in and repeated my question, "Do you guys have any spare cardboard in there we could have?"

This time, my request was met with a smile, "Sure, man," they said. I gleefully ran back over to Joey with our cardboard in hand. We had picked up a thick sharpie for just this moment and I fished it out of my bag. 'COLUMBIA' we wrote on both of our signs. I smiled, I felt as if we had just cracked the code to hitchhiking. How could we lose with cardboard to aid us now?

We walked for another two miles through some thick swampland that was littered with every sort of trash you could imagine; burger king wrappers, plus-sized porn CDs, trash bags, toilet seats, you name it and it was there. A sad sight that really represented the state of the human race to me. Here we were, in one of the most beautiful and unique swamps I had ever been to, probably filled with life of all kinds, but we use it as a dump because you won't get caught throwing your trash in it. How demoralizing.

We came out of the swamp to a crossroads and plopped down again, the sun was really taking it out of us and the signs weren't helping yet! Not that it was much of a surprise, just a little hope to get us moving. I looked over at Joey, he was not happy and just as tired, if not more so than me. I had to do something to get us back in the zone because we were losing it, fast.

Showtime

My greatest strategy for when I get tired and exhausted is to act ridiculously excited and eccentric so I forget how tired I am. I finally got to the point where I had to use my excited stage at our little crossroads. I shook off my backpack and stood up, the number one problem we had was that cars weren't paying attention to us, well, I could change that. I had a bit of a background of flipping and some minor gymnastics so I had a plan for the next car to come. "Joey, I got us, man. Watch this." I eyed a white car coming around the bend, first customer for circus ole featuring the one and only, Jake Blue.

I summed all the energy I had left and started a little jig, a little eye candy to make sure they couldn't miss the main course coming up. They were getting closer, it was time. I prepped all the energy and directed it into my legs, *Now or never,* I thought. I sprung myself up into the air into a backflip, I realized that I wasn't going to make it all the way around halfway through my rotation. "Well, shit." My hands hit the ground first so I didn't break my face on the rocks next to the road, but my shins hit next, and that hurt. I winced and caught a glimpse of the girl in the car giggling as she flew by us, spraying a couple bits of gravel in my face on the way by.

"That did not go according to plan," I muttered. I could feel Joey's eyes on me though, we needed this burst of energy ASAP. I popped back up to my feet, ignoring the stab of pain in my shins, and forced a fake laugh in Joey's direction. *I gotta keep going,* I told myself. I looked up to see our second customer in a red car flying down the road. I started my jig again, this time I

36

threw in a little Charleston for an extra flair. *Okay, they're watching,* I thought. *Now just don't fuck this one up.* I threw myself into the air again, this time a little harder despite the protest of my sore legs, this time I landed in froggy position and sprung into starfish pose with a big smile to make sure they knew that flip was for them. Vrraaaaaaooooooo, the red car flew by without a second thought about picking us up.

I couldn't help but smile as I thought to Russell Crowe in Gladiator. 'Are you not entertained?' Hehe. I was starting to feel a little spark light the coals in my heart. Car number three got a better performance, this time I landed on my feet with a smile. I was coming to life even though our third customer flew by me as well, but it was too late for them to bring me down, this was my time. It wasn't long before I was flinging myself into all sorts of cartwheels, flips, and other shenanigans on the side of the road. "Bet you never seen a hitchhiker like me bitches...LOL." I even walked back onto the grass to try a gainer (a running forward backflip) and landed directly on my face. That one really hurt, I dug myself up and brushed a stubborn clump of grass off my cheek. I was starting to think that maybe this high energy thing wasn't my best move until I heard Joey start to laugh.

I looked up to see him pointing at me, "You still have grass on your face, dumbass," he said. I smiled a real smile at him this time, *Mission accomplished, mood restored.*

I walked over to my brother in arms and offered him my hand. "Come on, dude, let's make it to some shade before we take a long break." He took my hand with an almost inaudible, "Fuuuuccccccckkkk," and got to his feet with a grin. We had a long, long way to go. Our mood lasted for another mile or so before we were back in the trenches, struggling for every step. "Just have... to keep... going." We both knew that if we stopped again, we might not get back up.

My Downfall

For 10 more miles, Joey and I limped, walked, and trudged for every step of the way. Everything hurt: our shoulders, our chapped lips, our sun-burned hands, but most of all, our feet were still just dying. Joey was coping by becoming more and more vocal, talking, cursing, he even pulled out his phone to play music and waste his precious battery. That ticked me off real good, I was using the birdsongs to keep me walking and now Joey had silenced them with his classic rock and singing. The importance of keeping us both in a good mood was not to be toppled by my need for Joey to stay quiet, however. I bit my tongue, and invited a very old side of me to come out I had not seen in a long time.

Throughout all of high school, I was always the good kid. I never drank, never smoked, I spent my mornings picking up trash on the school grounds. My good behavior was rewarded by my teachers and my friend's parents. I was a little angel, 'So well tempered,' they would say. What they didn't know was that all of my good behavior was only a side effect of my dirty little habit of basketball. I got angry and emotional way more than the average kid, I just had a good method of letting it all out. More basketball. I would break into the gym to shoot baskets in the dark just to let out all of the emotion of the typical school day. Those dark gyms were where I let out my real personality. The person that the rest of the world got to see was a fake, very little realness slipped through the shell I created off the basketball court. It was only after my injury senior year that I was able to break my shell and kill the fake personality that I had pretended to be for so many years in high school. At least, I thought he was dead until that long, lonely stretch of highway in South Carolina brought him back to life.

Joey took to calling him 'Focused Jake' on our trip, I feel that even calling him Jake was a bit of a stretch, for the personality I usually exhibit is fun, happy, and social. Focused Jake was characterized by long periods of silence, very little unnecessary movement, and a complete shutout of everything else so he could focus on nothing but the road in front of him. I wouldn't quite call him a split personality, but more of a coping mechanism to deal with all the stresses of traveling on the road and the throbbing of my infected cut from the glass the day before. Something important to remember, Jake and Joey are great friends, brothers even. Focused Jake and Joey? We could call their relationship something like turtles and dogs. Me the turtle, Joey the dog. Happy Jake tucks into his shell when Joey starts pissing him off. Happy go lucky dog runs over to try to get Jake to come out of his shell by playing with him, but of course, the more you try to coax a turtle out, the tighter it tucks itself in.

I could tell Joey felt something was off, I was usually the one injecting the energy into our glum situations, but I had shut down. It was the only way I knew how to cope with the pain of the road. Joey had no idea of the snap change that had happened so, being the great friend, he is, he tried to repay the favor of cheering me up. 'Hey, Jake, what happened to that girl you were seeing before we left?' 'Hey Jake, what kind of tree is this?' etc. I didn't have the energy to try to explain to him what I was feeling, so I used the only card I had left and picked up my pace.

The only way I could keep Joey from trying to open up my turtle shell was to make him struggle to keep up so he had something else to focus on. Looking back on that decision, it probably wasn't my best move. Shutting Joey out led to an infinite number of problems down the road.

The Power of Kindness

10 miles of miserable pain and walking later, we were just about to give up. As fate would have it though, just as we were ready to give up and sit down, a van pulled up to us on the side of the road. "Thank the heavens above," we were saved. There was no energy left for us to run to this van though, we limped as fast as we could to meet a man named Bill, a former postmaster.

Bill was also a former hitchhiker (no surprise there), but he didn't pick us up out of sympathy like the others had done before him. Bill picked us up because he desperately needed someone to talk to. As we got to know him, we soon found that his wife had left him, he very rarely talked to his kids, and being retired just wasn't good for him. That weekend, he was headed up to the lake house where him and his wife had spent so many summers before they were separated so he could, "Just remember what it was like." The more we talked to him, the more we realized that this was a man that was on the brink of suicide, his entire life had gone down the toilet and now he was left with no motivation to do much of anything. Everything he loved was gone.

Joey and I both unconsciously picked up on the signals he was giving off and listened to everything he had to say. For the next 40+ miles, Bill told us his entire life story and took us all the way to Lexington, SC. Over the course of the car ride, we spent a long time on the topic of death and what each of us felt about it. Joey and I went first so we could make Bill feel more comfortable opening up. Joey started us off with his view that he didn't care much about death so you could do whatever to his body after he dies. Flush him down the toilet, feed him to the Loch Ness Monster, both, he didn't care much. That got a little grin out of Bill. My turn to bring it home.

I said, "For me, I feel like when I get to 80-85 and it starts to get hard to get out of bed, this is what I'll do. Run down to the supermarket, buy a massive necklace of steaks, and drench myself in barbecue sauce. Then I'll walk out into the woods with a knife, find a good rock to sit on and wait. It might be a while, but we all know that a big ole bear is gonna come walking out of the forest looking for a meal, 99.8% chance I give it the best meal of its life and my soul is carried on in nature. Orrrrrr, 0.2% chance an 85-year-old man fights off a bear with a knife and goes down in history as the most badass grandpa ever!" That finally got Bill to let out his first laugh of the ride.

The time had come for us to get out as we arrived at one of the legendary Waffle Houses of the South. Bill pulled over for us and his sad demeanor returned to him. We shook his hand and thanked him the best we could, but a sad man thinking about suicide still drove away from us that day. There is no way for us to know what happened to that nice old man, we can only hope that we made enough of a difference to give him a reason to live.

Waffle House

We had been hitchhiking in the South for about a day and a half, not including our hotel night, and had already seen about a billion Waffle Houses. They popped out of nowhere and everywhere, which left us with only one option, we had to try it. Bill dropped us off in the parking lot in front of the Waffle House and Focused Jake had left to leave two happy kids laughing there in front of this mystery that was this restaurant. "Okay… let's do it, I guess," I said suspiciously.

Like walking into any establishment on this trip, we got some funny looks. The servers inside had just seen us jump out of the car outside with our cardboard signs in hand, they knew exactly what we were doing. When we walked in the door, we couldn't help but grin, everybody in the Waffle House had their red flags up from the moment we came in the door. It was almost fun to have that kind of impact on someone. To walk in and have everyone react to you is usually a luxury reserved for the extremely rich and famous. It was only day two, however, we were slowly starting to understand that people's reaction for us was not respect or even curiosity, it was fear.

The red flags made it hard to interact with anybody. It was easy to tell that nobody wanted to talk to us, nobody wanted us in their restaurant. Some people would crumble under that kind of treatment, but we were not at that point yet. Maybe it was the naivety, but I could not let someone not want to talk to me. My thoughts were, 'If anyone is going to not want to talk to me, they are going to have to tell me to my face while I'm asking them about if they believe in sasquatch or not.' (P.S. I learned it's called the skunk ape in the South.)

We found a spot in the far corner booth and watched everyone try to avoid eye contact with us. *O screw this,* I thought. I stood up and walked up to the counter with a smile and started talking to a lady about something or another to do with waffles. It wasn't but a minute before the whole restaurant was coming over to our table to talk to us about why on earth we had decided to hitchhike. This is where Joey came in, I had the goods that drew people in, Joey could sustain and seal the relationship.

We had a good couple of laughs before we realized we still had to order! Joey tracked down some menus for us to get our first look at the infamous Waffle House menu. The name would lead you to believe that the place serves mostly waffles, this was not the case as there were only about three waffle dishes on the menu. "Hmmmmmm, okay, I guess we'll just try for calories." I ended up getting some sort of an omelet and Joey got a ham and cheese bowl that would lead us to name Waffle House the much more appropriately 'Grease House.' Joey's food had standing grease in the bowl and my omelet looked like it was on a slip and slide with all the grease covering my plate. "Well, I guess we asked for calories," I thought, then dug in.

We finished off our meals like a couple of ravenous wolves, scarfing down everything they would give us. That amount of grease on an empty stomach is a killer though, about 10 seconds after finishing our food, we felt a gurgle in our bellies that warned us of unpleasant things to come. As we left the dreaded building, we made a promise to never, EVER come back. As a final send off, Joey turned around with both middle fingers in the air and proclaimed, "Fuck you, Waffle House!"

The Things Pizza Can Do

All jokes aside, we still had six miles to the Metzger's house (my mom's cousin's house). The irony of that was that as the crow flies, they were less than half a mile away just across the freeway. Being that we had no vehicle to cross the eight lanes of cars without ending up as roadkill, we were forced to take the long way around over the next exit. We both knew that we had used up our two ride quota for the day already, this was going to be another long walk. We didn't have much motivation to keep going after our previous 13ish miles walking in the heat, but as if on cue, my phone started ringing. I dug through my bag to find it before it stopped to see the caller ID read Bobby Metzger. I answered the cousin I had never met, exclaiming:

Me: Bobby, is this you?

Bobby: Haha it sure is! I understand you two are almost to Lexington. Is that right?"

Me: God, I hope so, man, if Google is lying to us, I might dig my own grave outside this Waffle House and die. It says we're only six miles from your house though! Can you still take us in tonight?"

Bobby: Of course, man! I'll be home around six with some pizza, can you guys make it?

That ended up being the magic word we needed to hear, my mouth salivated and I mouthed, 'Pizza' at Joey as if it were a mythical prophecy that had come true. I could see the switch flip in Joey as it did in me, nothing was going to keep us from the delicious, warm, melted cheese of a pizza!

Me: Oh, Bobby, I can promise you that we will be there.

Bobby: Ok, looking for… *click*

There was no time for more talk, Joey and I were on a mission to get six more miles in three hours. Easy, right? "Anybody can do that!" we exclaimed. The primal hunter mindset came alive deep within us, "MUST… HAVE… PIZZA!" we screamed with a battle cry that would have put even the greatest warlords to shame. Our legs spurred into action with energy we had not felt since our plane touched down in Charleston, nothing was going to stop us from getting to Bobby's house as long as there was food on the line.

Our pizza high lasted for five of those six miles before we had to switch over to the reserve tank. I don't think Joey or I have ever been that tired in our lives. We sat down for one last break to laugh at how beat we were. It was only five miles and we were toasted. Our backpack straps had started to cut into our shoulders and hips leaving us with long, linear rashes. The only bonus to that was that it looked funny enough for us to laugh at each other for a couple minutes before we went silent. We could both feel that it was time to finish strong, I could see Joey's chest rising up and down as if he was trying to breathe in the will to finish that last mile. I didn't blame him, I looked down to see my chest doing the same thing unconsciously. I looked back at Joey and found his eyes now looking at me. I could see the pain he was feeling, though he refused to say anything about it, I could see the weakness melting away from his eyes as the pain continued. We were going to be some tough mother fuckers by the time we finished this trip.

In a moment of self-reflection, I thought about our trip in a new light. *How weak are we that it is only three days since our plane landed in Charleston, yet we can be in this much pain?* Holding Joey's stare, I looked a little bit deeper and found something else; amusement. Amusement at our weakness and joy that it was quickly being purged from us the longer we didn't have enough food and the longer we forced ourselves to walk on our cut up and blistered feet. We were leaving the weakness we had been taught our entire lives behind for a new and better version of ourselves. My pained and hopeless look curled into a smile, at the same time I saw the wrinkles under Joey's eyes signal that he was about to do the same. Pure joy broke out between us as we reached over to grab one another's forearms, the ultimate brothers in arms embrace.

In that moment, our perspectives took a complete shift. We all know the saying, 'It's just the tip of the iceberg.' We've all heard that our entire lives, but until the moment that you are put in a situation far enough outside your comfort zone, you can never see the rest of the iceberg. Our eyes had been opened, and a new level of toughness came out of us that I had never known. We sat down on that street corner as boys that had no idea how we were going to keep going, and stood up as the beginnings of the men we wanted to be. We cruised through our last mile, not with ease, but with joy that we had to struggle so hard to finish. Turning our last corner there we saw it, Bobby and Keziah Metzger's house.

Family Time

Let it be known that before my mom had called her cousin, Bobby, a few months ago to tell him we were doing this trip, Bobby and Keziah Metzger had only met me once as a child. I was a stranger in every sense except for that I was distantly related through my mother to them. They didn't owe me or Joey anything. Neither Joey, I, or the Metzgers had any idea what to expect.

Joey was ahead of me as we turned our last corner into the Metzgers neighborhood. I found him waiting in the driveway looking up at a tall narrow house. I walked up next to him breathing heavily, "Care to do the honors?"

Joey laughed, "It's your family, bud." *Okay, that's a fair point,* I thought. There was a car in the driveway that signified someone was home. We made our way up the narrow driveway cautiously. So far, we had been led to believe that private property was viciously protected in the South, we didn't know what would happen if we had approached the wrong house. Standing at the front door, we started thinking. As soon as we pushed that doorbell, we were going to set in motion a whole different set of events to see what Fate had in store for us. We were pretty sure it was going to be good, but there was strength in being free of comfort with no one to take care of us. Before I could think about it too much, I pressed the doorbell and readied myself to run if it was the wrong house. A woman I didn't recognize opened the door, we watched her face screw into shock when she saw how ragged we looked. Dirt covering our faces, shirts drenched with sweat, we weren't the most appealing guest to anyone. "Hi!" I quickly said and extended my hand.

"My name is Jake and this is Joey, are we at the right house? Bobby invited us." The lady looked at us for an extra second with suspicion or amusement, I couldn't decide which. In that brief moment, it briefly crossed my mind that maybe Bobby had not told whoever this was that he had invited two hitchhikers over to stay the night. Just as I was ready to turn and try the house across the street, the lady stepped back and opened the door to us.

"So, you're the hitchhikers, come in."

Now I was suspicious, I could not read this lady at all. She didn't shake my hand, she hadn't said who she was, she did welcome us in. My mind flashed ahead to all the possibilities; was she a vampire? Did she have a pet alligator she was going to feed us to? Witch doctor looking to sacrifice us to the swamp gods? I had no idea what to think. I felt Joey make a move towards the door behind me and reassure me enough that if she was gonna turn me into a vampire, at least Joey and I would be vampires together.

We squeezed our way through the tight door and stood in this woman's living room. I think it was just a little awkward for her to let strangers into her house, but she approached us and introduced herself as Keziah (pronounced Kesha). We shook her hand with a huge smile on our face, it felt so good to be taken in by someone that was not going to flip us off or tell us to get off the road. "Bobby will be home at six, if you want some food, feel free to ask, annnnnnnd our shower is down that hall," she hinted. I leaned over to Joey *sniff sniff*. I don't think I have ever creeped him out more than the moment I sniffed his shirt for him, he looked down at me a little amused.

"What cha doing down there, bud?"

I looked up, "Yeah, I guess this is a little weird, isn't it?" I composed myself and caught Keziah looking at us sideways. I imagine she was thinking something along the lines of, *O good lord, what did I just let into my house?*

Joey went first to the shower, leaving me to entertain our host. I left my bag on the back porch and came back inside to get the scoop on every bit of high school drama Keziah remembered. It went something along the lines of, "Hey, Keizah, how many boyfriends have you had?" *Death Stare* "Ok! Are the kids home?"

She smiled, "Yes, they're right upstairs, I think Nolan is taking a nap." I was gone and up the stairs before she said another word to go wake up my cousin…or second cousin…or cousin once removed, not quite sure how that works.

I knew his name was Nolan and that was about it. That didn't stop me from running upstairs screaming "NOLLLAANNNN!" I didn't care if he didn't know the weird hitchhiker that was coming up to give him a hug, I had already decided we were going to be friends. *Knock knock knock* "Yo, you up yet? This is your cousin (or something along those lines) and I need a hug!"

All I heard was, "Huh? Mmmm nmm nmm." Alright, so that wasn't going to work, I guess I would have to talk about something that wasn't high school boyfriends for the next couple minutes with Keziah. It wasn't going to be easy, but for my family? I can do anything.

It took a minute, but we got the matriarch of the house to warm up to us after an hour or so (I think it helped that we both showered). We were laughing and had a great time meeting their kids, Maddie 16, Nolan 14, and Grayson 2. However, talking to the kids was much different from talking to Keziah. The kids were laughing about how crazy we were, but Keziah was different. I asked her something along the lines of, "You wanna join us for the rest of the trip?"

I set myself up to be made fun of on that question and she replied, "Not a chance," but she didn't attack us like most people did.

Joey and I had been made fun of for over a year leading up to this trip and had become desensitized to a myriad of insults. It didn't hurt our feelings, but we were pretty well versed on how people responded when we told them we were hiking across the country.

"Are you messing with me?"

"No, you're not actually doing that!"

"O god, you guys are so dead, don't you know people don't hitchhike anymore?"

Those responses were why Keziah's answer stood out to us. Just 'no' with a little grin. What was that answer? At first, I thought it was just an answer without an opinion, but the more I listened to her, the more I realized that a lack of opinion wasn't what made her answer. What was it about the answer that was so curious to me? Then a man burst through the front door with a different air about him than anyone I had ever met before. "Are they here?!" Bobby Metzger yelled excitedly. I didn't know him or even shook his hand yet, but it didn't matter, I already knew we were going to be friends.

It didn't take long before Joey and I were having the time of our lives laughing with the Metzgers late into the night. Bobby was what I would call a true family man, but far too unique to sort into any generalization. Bobby is the kind of man that you want to know everything about as soon as you talk to him. I call that sort of social gift the X-factor. We had the unconscious need to be his friend instantly... It is very hard to ask him the right questions to understand what it is that makes him who he is, but being so curious, we had to try. Joey would ask, "So, you grew up in South Carolina?" Bobby would answer all the questions, usually accompanied with some hilarious story, but then you would realize you still don't know what it is that makes him the way he is because we were laughing so much.

This went on for a while, we were having a great conversation, but I was utterly puzzled why I suddenly felt so close to these people. Then I caught the first part of my answer in looking at Keziah for a minute while Bobby was telling his story. She had lit up like a candle; there was almost a visible brightness around both Bobby and her that got brighter the longer we talked. There is a great saying that goes, "Behind every great man, is a great woman." Now, I don't take this saying in a way that a woman is always behind a man in any sense. What it means is that something about good positive feminine energy brings the best out of a man, I think it goes the same way for women. Bobby and Keziah had the most magical presence when they were telling their stories together and I think that saying perfectly describes it. It was like watching young love despite them having been together for years and years; they were truly bringing the best out of each other in that moment.

Nolan and Grayson retired after a while, leaving just Maddie, Bobby, Keziah, Joey, and I around the table. As the oldest, Maddie was permitted to join us and talk about the realities of the road. We told them about the small shot of racism we had tasted and our opinions on how ridiculous the constant fear we had been experiencing was. Bobby was enthralled to hear our stories; he has a deep-rooted adventurer in him that half-way wished that he could join us. Joey and I partly wished he could come as well, but one thing was sure, we had to leave tomorrow morning. We could not be sucked into the comfort of sleeping indoors and having people that love us around at all times. Joey and I had just made the mental breakthrough realizing that the trip is doable. If we kept going, we were going to be home in no time.

Begrudgingly, we gave everybody hugs goodnight and got ready to leave early morning. Joey took the deck outside to have some peace and quiet while I was offered a couch inside. I thought that it might be nice to sleep on something soft, so I turned the lights out and felt the squishy couch envelope me. I spent about 10 minutes wrestling with the pillows and cushions of the couch before I rolled off of it onto the floor, it just felt so unnatural to lay on something soft. We had only spent a single night outside and already I was adapting to Joey and I's new lifestyle.

Time to Leave

We woke up around 6 a.m. feeling ready to go conquer the road now that the Metzgers had restored our faith in humanity. Bobby had said he would take us as far as he could towards Georgia before he had to leave to work. At 6:45, we hopped in his car and started flying across the South Carolina highways, a pleasant change from the almost 20 miles we had walked yesterday. Bobby was driving like a mad man in a hurry to help us as much as he could before he had to go to work. About 30 miles later, he pulled off to the side of the road, the vibe in the car changed from enjoyment to somberness. After hugging us both, Bobby hopped back in his truck and rolled down his passenger window to wave goodbye to us. Right before he left, I saw a longing in his eyes, he so badly wished that he could drop everything and just start walking with us. There was a moment of hesitation in him where I could see him, just for a second, considering what would happen if he came with. He swallowed his longing and with a final wave, pulled back out into traffic and left us on the side of the road next to an old church.

Watching Bobby drive away left a bit of a hole in Joey and I's hearts. We were going to miss that family, they reminded us that no matter how mean the people driving by us were, there were always good people out there that would lend a helping hand. We pushed down our sadness and turned back to the west. Next destination, Joey's long-lost aunt in Atlanta, Georgia.

Clueless

Making it to Lexington had been a huge step in showing ourselves that this trip was possible. Atlanta was going to be a whole different beast though. Charleston to Lexington had been about 120 miles and we managed to do it in two days, which was a huge boost of confidence for us. However, Lexington to Atlanta was over 200 miles by foot and was more densely populated. What made me most nervous was that we were progressing deeper and deeper into the South, the home of American racism. We had no time to worry about the what-ifs of our trip though. We were out here to conquer fears instead of letting them continue to dictate our decisions. I knew the next leg of our trip was going to be hard, but I did not know how much it would alter the way I look at the world.

Our Illogical Hopes?

Joey and I contemplated for a couple minutes about what to write on our signs, eventually deciding upon Athens, Georgia. One of our rides had told us it was a must-visit destination and we were very excited to go. Now, we just had to figure out what people wanted to see on our signs so they would pick us up. It ended up being very fun to experiment with what people would respond to on our signs. They had to be big enough letters that our potential rides would be able to read them, but also innocent enough that we didn't come off as serial killers that were waiting for victims. You could tell how our day was going from what we would put on those signs. They ranged from *What would your God Do? / Will Sing and Dance for Rides! / Pick us up or Fight us. / Traveling Country via Generosity! / I'd Pick You Up… / Ladies! Your truck bed is good enough!* We would also include all sort of hearts or dinosaurs over our signs to spice them up a bit, but my favorite sign that didn't work at all said:

Athens
College?
Nice?
Cute?

Even after all the time of experimenting with different signs on the road in different states all across the country, we still have absolutely no idea if signs are beneficial, detrimental, or have no effect on us getting picked up. The only thing that seemed to be related to us being picked up was whether or not the people driving by were former hitchhikers that could sympathize with us. Everything else we did, whether it be sitting still, holding dance/parkour parties on the side of the road, ballroom dancing with each other, climbing street signs, yelling and chasing cars that wouldn't pick us up all seemed to have absolutely no effect on if we got picked up or not. Just for example, I sat for around 90 minutes with my thumb up on an exit with heavy traffic. The second I gave up and started polishing my knife, a lady stopped and picked me up.

There did come a point when we started to understand that no matter what we did, people were still probably going to drive by. The hopeful side of us didn't want to believe in that though. We had hope that we could somewhat control our situation if we could just improve our signs to the point of people that would want to pick us up. Even if our roadside behaviors had no effect on drivers, it did help Joey and I stay sane by believing that if we got better, the people around us would also get better.

Homesick

It didn't take much walking after Bobby dropped us off to get picked up by another former hitchhiker (sensing a pattern here?). The man said he could take us a couple miles down the road, but it wasn't going to be much. As always, Joey and I accepted and thanked him, but the rides were starting to blend together for me. Focused Jake was taking more control, leaving Joey on his own more and more often. The exact opposite of what we needed if we were going to survive this trip that would soon prove to be dangerous.

The man dropped us off at his driveway, eight miles closer to our destination. I will always remember that first stop as the one and only time I had true heartache for home. I pulled out a carrot for each of us and the homesickness hit me like a ton of bricks. Looking down at my half-eaten carrot, the thought of all the people back home that I had made connections with suddenly pulled on my heartstrings. Morgan Stark was one of the most beautiful people I had ever met and she loved carrots. That led me to think about Emily Peinado, who was one of Morgan's best friends that I had grown to adore. I took the bait of those first thoughts and ran down the rabbit trail of thinking about what my family would think if I didn't come back, *What about all the money they've invested in my education?* I asked myself. I thought about how I had unintentionally left my brother at home after my parents separated by going to college and left him to all sorts of potential drama. I would never be able to make up for that if I don't come back. My friends that I had promised to travel to Mexico on motorcycles with would take that trip in sorrow because I couldn't be there with them! Suddenly, I snapped back to the present, that line of thought existed in a mythical time and place that I had created in my head, and it was far from reality. I was not going to let the fake emotions of something that hadn't happened get me down, I had my brother in arms next to me, that is all I needed.

A Day of Surprises

The only moment I ever felt homesick had come and gone, I was going to be okay as long as I could keep my mind in the present. Restarting our walk, we got the feeling that we were not going to be running into anymore rides for a while and, of course, we were right. The only good news we had found in our form of travel was that we could count on some new experiences that would spice up our never ending walking every day. We had made it to the next town and filled up on water, but just as we were leaving city limits, we saw a big white suburban looking vehicle creep past us. Joey and I looked at each other from across the road, his face told me he was thinking, *That's suspicious,* just like I was. The big white car rolled up the hill for a ways before pulling off in a turn out. Okay, this was weird, whatever they wanted had something to do with us.

Watching the car carefully, we kept working our way up the hill on each side of the road. When we got within a couple hundred feet of the car, two cops opened the doors and leaned on the white suburban waiting for us. Having them be cops was somewhat reassuring, but now Joey and I were more than a little curious. Why did they stop? It didn't look like they were wanting to give us a ride because they were not smiling, there was something strange going on. Finally approaching the car, I shouted a greeting, "How's the heat treating you gentlemen?" One of the cops half smiled at our comment and motioned for both of us to join him over by his car.

Joey and I had always been friends with the cops where we were from, so we were happy to oblige and follow their orders. We shook the cops hands and introduced ourselves as Jake and Joey, "How may we help you gentlemen?" I asked. You could see a mixture of utter confusion and trying not to laugh written on their face.

"Uhuh…what are you two boys doing walking on the side of the road?" they asked. Joey and I looked at each other, it was our turn to be amused because we knew they wouldn't believe us.

Joey responded with a smirk, "Well, we're hitchhiking from Charleston back to Idaho." Total confusion was now written across their face again. They looked at each other, I could tell they were trying not to laugh.

"Ummmm okay…why are you doing this?" they asked.

I had to keep myself from laughing to answer, "Gentlemen, we've lived privileged lives and thought it was time to come out and suffer a bit." I let my words sink in and realized how little sense that probably made to these people. That was about the worst way I could have just explained that. We spent the next couple minutes telling them every detail of our trip so far to convince them that we really were hitchhikers from Idaho. Before they left, they shared an important tidbit of information. We asked them why they had come to talk to us and they said, "People call the cops if the wind blows the wrong way around here, we won't be the last cops you see." Joey and I looked at each other and agreed that we were more than happy to see the cops as long as they gave us a ride next time.

That was not the last new thing that would happen to us today though. A couple of miles later, a lady stopped us by yelling off of her porch on the side of the street. "What on Earth is y'alls doing walking in this heat?" We turned with a smile, nothing like a friendly southerner to put some hop in our step! Joey was closest to her house, so he yelled back, "We're hitchhikers! Going back to Idaho!"

She laughed, "Iowa is a long ways away, boys!" We had to take a minute to laugh this time. Is there not a single person east of the Mississippi that knows where Idaho is?

Joey: No, we're from Idaho! By Washington State!

Southern Lady: *Laughing hysterically* How… the hell… do y'all think you're going to get all the way back there?

Jake: Traveling via generosity, ma'am, just seeing if it's possible!

Her toddlers playing on the porch had taken interest in us. One of them made a move to walk down the stairs of the porch towards us, but was quickly stopped by his mother who stepped in front of him. Joey and I had seen enough fear on this trip to know this lady was scared of us after that move. She didn't want her kids anywhere near us. Not wanting to cause any trouble, we turned to continue walking but we were stopped by another yell from this lady standing on her porch.

Southern Lady: Hey, wait!

We turned back to her with a smile, happy to oblige any questions she had, but things started to get strange. You could see the mental cogs turning as she struggled to find anything to ask us about our trip even though fear was written across her face.

Southern Lady: Where are y'all coming from?

Joey: Charleston, ma'am! Made it about 140 or so miles so far!

I went to add something on to Joey's statement, but I was tired of yelling, so I made a move to walk up the lady's driveway. Quickly, she stopped me.

Southern Lady: You can stay right there!

Okay, things were starting to get really weird now. What was it she wanted from us? She clearly didn't want to talk to us that bad but insisted that we did not leave. I caught a glimpse of someone on the phone through the window and started to get a little suspicious.

Me: I'm sorry, ma'am, I didn't mean to intrude. Is everything okay?

Southern Lady: Yes, but I can hear you just fine from there!

We fluff talked for a couple of more minutes before she finally gave us a tiny piece of information that told us the story.

Southern Lady: Y'all know there was a prison break just down the street?

Me: No, what happened?!

Southern Lady: Three men escaped from prison, it's all over the news! Cops are going up and down the streets looking for 'um!

It all clicked into place now. She thought we were the escaped men from the prison, that's why she is so scared! Joey nudged me on the shoulder, "We should go, Jake." My mind flashed back to the person I had seen on the phone inside. *Shit,* I thought. She already called the cops and is trying to keep us here until they show up. Joey was right, we needed to go. The last thing we wanted to look like we were doing is harassing a black woman and threatening to come onto their property. Who's going to buy the word of a hitchhiker if it comes to our word against hers?

Did She, or Didn't She?

We politely said goodbye to the lady and went on our way, not turning around when she yelled at us this time. I turned to Joey to tell him they had probably called the cops, but he was already 10 steps ahead of me. "Jake, they called the cops on us, man, we don't want to look like we're harassing anybody when they get here." I had my mouth open to ask if he was thinking the same thing, but he had said everything I had on the tip of my tongue. I closed my mouth and nodded, I had faith that Joey and I could talk our way out of just about any situation. Being the eternal optimist, I said, "At least if we go to jail, we'll get a free ride!"

Joey laughed and looked behind us, "Yeah, unless the jail is that way." I turned and looked back down the long highway we had just walked over, heat waves rippled through the air and radiated off the road behind us.

"I really hope you're not right, man."

Another couple miles and amazingly, we had still not been pulled over. We didn't know what to think about that lady. *Maybe she didn't call the cops? Maybe she did and they hadn't found us yet?* We weren't particularly scared of being pulled over, for we had done nothing wrong. The real fear was that we would be taken back to a jail in the town we had just walked out of, that would make this a very long day. We had no desire to walk back over this forsaken road with no cars in such heat! I kept turning my head to see if I could spot any lights coming to pick us up, but eventually I forgot about them. There was no way to know if they were coming to get us or not, and there was no use in looking over my shoulder for the next who knows how long, all we could do was keep walking and hope we got picked up soon.

How Do We Stay Us?

The heat was really starting to ramp up, it seemed like every turnout and open spot was calling our name, asking us to sit down for just a minute. I could feel the cut in my foot starting to throb again, I can't imagine Joey's feet felt much better. We had started off day three on the road so hopeful, but it seemed the more hope we had to be picked up, the less likely it was. It seemed like only after the road had sucked all of the hope out of us and we were ready to curl up and die in the ditch on the side of the side of the road, a car would putt around a bend in the road and stop for us. Though we didn't yet know it, these long, painful walks in the southern heat were teaching us a fascinating lesson about human psychology.

Everyone is a product of their environment, there are some deep genetic personality traits ingrained in us but, as me and my brother in arms learned, it takes a fortitude that neither Joey or I had to remain yourself in an environment that is asking you on a daily basis to change. We were constantly feeling a pull to be colder to people we met on our various stops. There were sometimes dozens of people each day that would flip us off, swerve at us, or swear out the window, 'Get off the road, dumbass!' We had dropped into a hostile world and when you find that the vast majority of people don't trust you, don't want you, and aren't even willing to give you a chance to change their mind, it is extremely difficult not to reciprocate their behavior back at them. If we rewind the clock to the dawn of our species, the pull to reciprocate the treatment Joey and I were receiving makes total sense.

Adaptability is one of the traits that makes the human race so wildly successful. If you looked at nothing but physical traits, it is hard to imagine that humans would rise to become the dominant species on earth. Compare us side by side with that of a polar bear. If you throw us in a ring together, it is not hard to imagine which one walks out alive. While our ancestors were slightly more formidable than us, with more hair on their bodies, more durable skeletal structure in their jaws and skull, stronger teeth and claws, it was largely the same story with them and the fauna that they knew as their neighbors. Mammoths that outweighed even the largest of African elephants, short-faced bears stood 12 ft high on two legs and, despite weighing approximately 1500lbs, there are estimates it could run well over 40 miles per hour. These are only two of the formidable creatures that roamed the ancient landscape that seem to vastly outclass our ancestors, yet here we stand. A testament to the survivability of the creatures that came before us. So, what is it that has made the human race so successful?

I'm sure we are all well versed in the answer, we were smarter. We were able to develop language that led to complex social hierarchies all thanks to that thing between your ears called the human brain. More than just being smarter is that we are also an extremely adaptable species. If you look to the earlier example of the polar bear vs human example, the polar bear would take a human out any day of the week physically, no questions asked. Let's look at this a bit deeper though, if you dropped off a polar bear in the LA basin and tried to teach it what it needed to do to live, you couldn't do it. The urban environment is too foreign, no matter how long you tried or what you did, you couldn't teach a polar bear to survive in LA. On the other hand, if you grabbed a person from the LA basin and taught them everything they needed to know to survive in the North Pole, there is a chance that the human could survive there despite it being totally different from anything he/she has ever known. This showcases the generalist aspect of humans combined with our extreme adaptability, it has allowed us to be able to not only survive but thrive on six of the seven continents on the planet. The reason we have been able to have such success has been because of our extraordinary ability to adapt to almost any situation and discover new ways to survive in them.

One could hardly imagine how this amazing characteristic of humans could be harmful to us, right? In Joey and I's case, the rapid adaptation could almost be considered a detriment to us because of the mental toll it was taking. We knew that by coming to the South, we were likely going to meet different kinds of people that weren't all that friendly. We wanted to be their sun and show them that whatever they did, we would reciprocate with love and happiness. It seems like a great plan, but it is a fundamental rejection of human evolutionary programing. We were feeling a pull to be colder and not open ourselves up to people. Every time someone was mean to us just because they could be (which was a lot), we had to consciously stop ourselves from reciprocating their behavior.

The Happy Few

The good news in all the anger we were receiving is that it seemed whenever we got to the point that we thought we just couldn't absorb any more hate and meaningless anger from those around us, we would run into the most wonderful people. That was exactly what happened on that 11th mile of the fourth day.

I remember the shop vividly, there was a large field on our left that ended at a huge blue building that looked like an auto body shop. We were running low on water and hadn't seen any other buildings in a long time, maybe they had water? Turning off the highway, we walked up the dirt road towards the blue building with little clue of what to expect. There was a man outside that looked like he had been working as hard as we had and that impressed me. It is easy to feel a kinship with others who are suffering the same way you are. Knowing that we would probably receive the same cold-shouldered treatment as most other people had given us, we summed up all the optimism we had left and asked the man if there was a place we could fill up our water. I watched his face as he turned towards us, seeing two hitchhikers practically dripping with sweat could only coax one expression it seemed. Shock crossed his face. *Predictable,* I thought. Now the mystery was what expression would come next, most people's faces would go into initial shock into fear, what would this man do? Surprising us, the man's shocked face melded into curiosity. He actually took a step towards us so he could tell us to follow him to the sink on the side of the building.

Wow, it is amazing what the tiniest acts of kindness can do for somebody. I could instantly feel a slight bounce in my step return as the man introduced himself as Edgar. From what we gathered, he was just about the best mechanic in the state as he told us that he had contracts with a couple different sheriff's departments to keep their cars running at peak performance. We were more than willing to take a break and listen to Edgar's stories, but he was quick to tell us he still had some work to do before he was done for the day, "It was nice talking to you boys and good luck in getting home," he said with a smile.

We smiled back at him, how could we not? He had exceeded every expectation we had for him and given us another shot of hope. Most of Joey and I's conversation had turned dark until Edgar breathed life back into us by giving us no more than five minutes of his day. We made our way back to the highway and continued walking, our conversation with more of a spark in it now. Not even a mile away from the shop, we were starting to work our way up a bend in the road when a rusty old pickup truck pulled up next to us. It was Edgar.

Edgar smiled and motioned us to come closer. Joey and I ran over to receive one of the most kind-hearted things anyone did for us during our time in the South. Edgar reached in his back seat and pulled out a plastic grocery bag full of snacks and gatorade, his lunch for the day. "I know you guys are having a hard time getting rides, so I wanted to give you a little energy boost," he said and offered us the bag. Joey and I were quick to realize that it was his lunch for the day and tried to refuse. "Hey, look at me," said Edgar. "If you two are gonna make it home, you gotta start saying yes when someone offers you some food. I don't need all the junk anyways," he winked. Before Joey or I could argue with him, we were interrupted by a dump truck flying around the blind corner that Edgar was stopped on! Edgar threw the snacks at us and took off as fast as his little car could before the dump truck took us all out. A couple seconds later, he drove back by us laughing, "Good luck, boys!" he yelled. We watched him drive up the hill we had just come down back to work.

Joey and I looked at each other emotionally, I actually felt a tear of happiness come to my eyes as we looked down to see all the amazing gifts we had just received. Cheesy crackers, red and blue gatorade, an assortment of granola bars, it was a hitchhiker's dream meal. We sat down on the side of the road to fill our empty stomachs. Now slightly less hungry, we drifted off with smiles on our face. That small act of kindness had started to turn our tough day into a very fun one.

Cop Party

'Whoop Whoop' Joey and I woke suddenly to sirens. A cop car had just driven by us. We watched as it slammed on its brakes and turned around, coming to a stop right in front of us. "Well, that's probably not a coincidence," said Joey with a grin. I grimaced and shook my head; leave it to Joey to make a joke the second we get pulled over. We sat and awaited whatever was to come, but the cop didn't get out of the car.

Me: Alright, should we go introduce ourselves?

Joey: Haha, I guess…he better give us a ride.

We left our bags behind and walked up to the big SUV that had its lights on. We stood in front of the car for a while but still, nothing happened. The windows were tinted enough so that we couldn't see anything inside. We had no idea what to think.

Joey: You think he's turning off all his cameras so he can run us over?

Me: God dammit, Joey.

That's all it took, we busted up laughing in full view of whoever was in the car watching us. Now we didn't only look like hobos, we were crazy drugged up hobos laughing at nothing. That was probably why the cop inside did what he did next. Slowly, we recomposed ourselves and I decided I was going to ask the cop in the car what was going on. I worked my way over to the driver's side window and went to knock. Right before my hand touched the glass, the window rolled down and I was met with an officer yelling, "Get in front of my car where the camera can see you!" as he unholstered his pistol. I put my hands up right away and walked back to the front of the car without saying a word. Joey looked a little taken back by my instant change in mood.

Joey: What happened?

Me: He drew his gun and told me to wait.

You know what Joey says to me after that? He didn't even take a moment to pause and understand that this situation had just gotten serious.

Joey: Heh heh, dumbass.

I was so shocked I didn't even know what to say, I looked over at Joey in total surprise. He made eye contact with me for about a second and started laughing his ass off! I tried to be mad but it was so funny that he was laughing at me for getting the cop to draw his gun that I couldn't help it, I started laughing too! Just as we stopped, another two cop cars pulled up and surrounded us on the side of the road. We looked at each other again, I promise you I tried so hard not to laugh, but I couldn't take it. This time, we laughed so hard that we were leaning on each other for support, I can't imagine what the cops were thinking. By the time we composed ourselves, six cops had shown up on scene! We had no idea what we did or why they were there, we only knew that whatever came next was going to be very interesting.

Finally, the doors on the cars opened and cops piled out, completely surrounding us. Joey and I aren't exactly the kind to be shy so we made a point to introduce ourselves to everyone. "Hi, my name's Jake, nice to meet you."

"Hello, my name is Jake, nice mustache by the way! What's your name?" If you were blind, you would have thought we were at neighborhood barbecue. Then the head sheriff walked through the ring of cops that had surrounded us and shook our hands.

Sheriff: We got a call two men with big backpacks were trespassing out here, would that be you?

Us: Uuuhhhhhhhh, maybe?

Me: Were you saying something, Joey?

Joey: I wasn't saying anything.

Us:

Me: Possibly.

Joey: Yeah, maybe.

Sheriff: … You were 'maybe' trespassing? Is that right?

Us: Uh…yes, sir.

e had no idea if we were trespassing or not, all we knew is that we walked up to the blue shop, but Edgar wouldn't have called the cops on us, right? The sheriff looked past us to our bags on the side of the road.

Sheriff: Those y'alls bags?

Us: Yes, sir.

Sheriff: Then yes, you were trespassing.

So, it turns out the power lines on the sides of the road are actually property lines in the South. We were technically sleeping in someone's yard again. Careful to leave out that we had already done this once on our trip, we explained to the officers that we were innocent hitchhikers with no idea what we were doing. That got a little smile from them, what had started as a potentially very dangerous situation, turned into the funniest thing I had ever been part of.

Just imagine six cop cars on the side of the road blocking one lane of traffic while creating an arena to bullshit with the cops for as long as we wanted. Maybe the giggles were just in the air, but we had all the cops laughing 10 minutes after they pulled up as we told them all the crazy shit that had happened to us. By the end of our storytelling session, we had made a bunch of new friends, everyone knew there was nothing to fear. We had to have half of the cops in the county there, it made us think about what was happening in the rest of the county while we were distracting these six. One of them must have thought the same thing, wiping a tear out of his eye from laughing, he said, "Okay, boys, you're free to go, good luck!"

Graciously, we thanked them and went to grab our bags, but the head sheriff shouted, "Hey! Before we go, I was wondering if you boys would want a ride?"

Our jaws dropped, "Uuuhhhh, hell yeah, we want a ride!"

Because each cop had shown up in their own car, there were front seats open. One cop took Joey and one took me for the next 21 miles, IN THE FRONT SEAT! We had gone from potential suspects to cops in training in under an hour. Letting us in the front seat really speaks lengths about how awesome and trusting these cops were, because there was every weapon imaginable within my reach when I got in. On the sheriff's right-hand side, there was an 870 mcs combat shotgun with the capability for slug, buck shot, and less lethal like taser shot, and bean bags. Over my head was some sort of M4 looking assault rifle with a clip already loaded, not to mention the glock and taser on the sheriff's belt, we were fully loaded. They must have loved us because Joey said the same thing for his car, weapons everywhere. I don't even think you can attribute that to any Southern hospitality, there was just something in the air that day.

The second most memorable ride of my life ended 20 minutes later thanks to the cops being able to go 70 on the 45 mph roads (will get to the number one most memorable ride in Wyoming later). Shaking my hand, head sheriff let me out to meet Joey who had just gone through the same thing in his car. I will always remember what Joey said right as they left, "That…was…fucking…awesome."

We started laughing again, "Can't argue with you there, brother," I said with a grin.

The Southern Forests

We were off to our best start of the trip so far. Bobby's early morning ride, along with our cop encounter, had already put us well over 50 miles. We were feeling pretty good about ourselves now that we were in a position to beat our previous record of 60 miles today if we dug in for another couple hours on foot. We took a long, well deserved rest under a little tree outside some sort of fast food restaurant and plugged in our phones into an awesome little solar charger we had brought on the trip. Being well rested and high on life after the cops, we got our workout on with some push-ups and shadow boxing while waiting for our phones to charge. Just under an hour later, we set out to shatter our old record.

Focused Jake was starting to set in much quicker and easier now with only momentary lapses of Happy Jake. Joey had eased off of me about it, partially because he was in a good mood after our good day so far, and partially because he probably just didn't want to deal with me. The only thing that was bringing me out of my alter ego was that we were starting to progress into the foothills of the Appalachian Mountains. The fields had been replaced with thick trees on either side of the road, which was thrilling to me!

My college major being in Ecological Restoration through the College of Forestry, I had learned about all sorts of Montana environments and the ecological factors that make them tick. Now I found myself in a new environment that I knew very little about. When we would take a break, I would dig up some soil and try to figure out how old it was and what sort of minerals it was composed of. I learned a lot about the South in my surface level investigations, but the number one thing I discovered was how strong the mark of human presence was. I saw two big impacts of humans while in the South. First, I saw that the extreme habitat fragmentation had left unfavorable environments for keystone animal species like burrowing tortoises and large cats like the cougars that used to roam this area. In other words, people being so spread out leaves little room for animals with specific habitat requirements to help ecosystems function. Second, I saw that fire ecology had been vastly altered by fire suppression tactics put in place by the federal government's anti-fire movements of the 1970s.

It made sense, Joey and I couldn't go more than 15 miles without running into a gas station. There were no natural barriers in the South like in the West. Back home, the Rocky Mountains' uninhabitable and diverse landscape prevents people from living in many areas, allowing us the luxury to let some fires burn. In the South, the land was mostly flat, which allows people to sprawl across the landscape for as far as the eye can see. If the historic fire regime of lots of low intensity fires was allowed to continue, whole towns could be in danger. Now, as a result of suppressing the fires for so long, it almost seems as if fire has left the landscape all together. The forest floors are covered with thick leaf litter that never burns; I can only start to imagine what ecological consequences there were to excluding fires from those forests. Pinecones that used to release seeds when exposed to heat no longer reproduce. This leads to changes in the primary vegetative species, and maybe even eliminate some tree species all together. Depending on the functioning role of those fire dependent tree or plant species, their decline in the wild could lead to a number of other animal and other vegetative declines from species that depend on them. Eventually altering the forest so much that the problems created by eliminating fire can no longer be fixed by allowing fire to burn again.

Record Breakers

We had passed our 60-mile goal, but we couldn't stop walking. Almost as soon as we started off again, we were picked up by a lovely Mormon woman that took us another 13 miles. She confessed to thinking that we were part of the church the way we were dressed and that is why she had picked us up. I was expecting her to not like us after we told her we were not religious, but she surprised us by saying that the Mormon Church had a lot of lonely ladies just waiting for a big strong hitchhiker to join them. We laughed her off and she continued to tell us stories about how her husband was starting some sort of music festival for the little towns in the area to come together. Joey and I were interested in this lady, she seemed to have an awesome story to tell, but she was far more interested in us than her husband's music festival.

She wanted to know about the lives we lived back home when we weren't hitchhiking, so we told her everything. I spun her stories of the woods and running into bears, and Joey talked about his complex astrophysics theories that he had come up with in the past few months. She ate it up. Earlier, we had told her we were looking for a possible place to sleep that night so she took us on a tour around a golf course and down to a bridge helping us look for anywhere that we thought would be good to sleep. Eventually, Joey and I decided we had to get out and start walking at some point so we told her that a bridge on a back road would be good. It's funny how the day before had been one of the hardest days of our lives and this day was one of the greatest days we had hitchhiking. It just goes to show that no matter how bad things get, there is good just around the corner.

We hung out at the bridge for a minute, deciding against jumping off into the murky waters that could hold any number of venomous snakes. We were over 70 miles now and it wasn't even getting dark yet. We couldn't stop just yet, but we now found ourselves with a decision to make. There was a state road that led to Athens, Georgia that we could walk on with a lot of cars, possibly a better chance of being picked up? Or there was a long lonely looking road that forked off into the forest, we knew they both lead to the same destination, which one should we take?

We debated for a moment and decided upon the mystery road even though we had hardly enough food to last us through the next day, this trip was all about exploration though! We simply couldn't take the same road every other visitor to the South takes. So, we walked off into the unknown on that little road in the forest, knowing that excitement would soon meet us there.

It ended up being an awesome decision that led to talks Joey or I never would have had if we had chosen the main road. Where we now found ourselves seemed like the one stretch of road that had been forgotten by the rest of the world. No more cars drove past us, we were totally alone for the first time on our trip. What an interesting pocket of isolation in a place filled with people. It was getting to be evening in the Southern forest we had been walking in for so many miles. Up ahead, we could see buildings which was good news, we were running very low on food. When we made our way up the hill, however, we discovered the buildings we had seen were the remnants of an abandoned town unlike anything we had ever seen before.

We took cover under the overhang of an old brick building and took note of how nature had reclaimed the building as her own. Birds' nests and bat guano told us that this was a night roost for many of the flying critters of the forest. Vines had grown through the windows into what seemed to be some sort of old library, and new growth had taken seed in every concrete crack in sight. The only thing that didn't look 100 years old was the road that ran through the middle of this ghost town. We took a long break in that little town to take in the feeling of holding history in our hands. We stayed there for well over an hour as the sun started to go down. It was time for bed.

I threw down my bag and unbuckled our tent. Joey, looking at me sideways walked over and asked me what I was doing. "It's time to go to bed so we can get a good start tomorrow," I told him.

"Jake, it's only like seven, man, we can get in another hour of walking easy," countered Joey. I was not a fan of his logic, but after a moment of brief arguing, he convinced me to pull myself up off the ground and continue, thank goodness he did. As the sun sank below the trees and the light got low enough that it was starting to get hard to see the road, a pair of headlights came over the hill behind us and stopped next to us in the dark. Another former hitchhiker and Vietnam veteran rolled down his window and invited us into his car.

Spirits?

This man was different than the others that had picked us up, a new breed if you will. He was a former sniper that had killed so many men that he refused to tell us a number because of the horror the number instilled in him. Although he wouldn't tell us the exact number, he did say he remembered every single time he ended someone's life and that he had not told anyone this story in 30 years. This man had picked us up because he was still trying to make up for what he felt to be a lifetime's worth of sins that he had committed in Vietnam. Our old veteran took a liking to Joey sitting in the front seat and talked to only him for most of the ride. Often only whispering the horrors of war as if that speaking too loud would bring his victim's ghosts back to haunt him.

Cigarette after cigarette, we putted along that back road, never exceeding 40 mph while this man confessed his sins to Joey. He took us through the next town down to a gated nature preserve and dropped us off in the dark. I was almost irrelevant at this point, but Joey and that man had formed a connection that went deep enough for the man to offer to keep driving us around until we found a nicer spot to sleep that night. Joey assured him that he had done more than enough for the day and shook the old man's hand. I looked over at Joey as the man drove away, "What did he tell you?" I asked. Joey's eyes turned misty as he recited the man's stories to me that I couldn't hear as we walked off into the night. Those stories might be what lead us to the bizarre experiences we found later in the night.

We had to put our headlamps on by the time we finally found a flat spot called 'The Sanctuary.' The place seemed spooky from the moment we walked in, Joey really didn't like it and felt like we were being watched. I couldn't disagree with him and definitely felt there was some sort of presence about the place, but I was curious about this new spot. There was something alluring about it despite the shadows that seemed to dance on the trees in our peripheral vision. We set up the tent and decided to investigate further into the shadows that danced in the night. We found a path that led into the woods and walked down it with our knives drawn.

Joey was creeped out more than I was, maybe he was more sensitive to these spiritual sorts of things than I was. I felt something as well though, it felt like we were being studied, but there was no way to know. The forest was so thick you couldn't see 10ft into it, even with the headlamps. I was ready to go to bed, but I trusted my partner's nerves with my life, so we walked until whatever had spooked Joey started to subside. The shadows had stopped moving in unnatural ways except for the occasional flicker out of the corner of our vision. That would have to be good enough for us to sleep tonight. Joey cooked a family sized can of beans (he still can't eat beans because of how much he ate that night) and I snacked on the last of my granola bars before we fell asleep.

What Are You?

I woke in the middle of the night breathing heavy, whatever it was I couldn't understand about Joey's earlier wariness, I now was alerted to. It just felt like we weren't alone, Joey was asleep but this time I didn't wake him, whatever I was feeling felt personal. I listened outside to the occasional scurry or bird cry while whatever the presence was had enveloped me like a blanket. I had the worst urge to turn on a light and go outside to see what this mystery was that haunted us. I went to reach for a flashlight but something stopped me, I didn't feel like being able to see in the blackness outside would answer any questions for me. Briefly, I considered waking Joey but rationalized that I just had to go pee, no reason to wake Joey over that. I unzipped our tent and wandered off into the dark.

I had seen too many snakes on this trip already to go too far barefoot in fear of getting bitten. I stopped not far from the tent and peered into the dark. The shadows were dancing again, but the allure to them had worn off. I had more important things to tend to as I unzipped my pants to do my business and watched them dance with impartial judgment. I didn't know what I could trust my mind to see in the night and didn't feel threatened by the shadows, so I let my imagination run wild. A couple minutes later, I found myself back in the tent with no more than a couple of mosquito bites. Joey grunted as I unzipped the tent, slightly aware that I was coming back. I grunted back to let him know everything was fine before I settled into sleep like a rock the rest of the night.

Getting Stronger

Day four on the road and we woke up feeling like champs. Only one pair of cars had driven into The Sanctuary last night (probably to do a drug deal), which left us to have the most peaceful sleep on the trip so far. I felt energized! Our bodies were getting harder and stronger by the day. We hopped out of the tent around 6:30 a.m. to brush our teeth and I caught a glimpse of Joey's abs ripple in the morning light. "Dammmmnnnn, Joey," I winked at him and pointed at his developing 8-pack.

He laughed and pointed back at my stomach, "We don't have any fat left, Jake," he pointed out.

Joey didn't seem near as thrilled as me about our new discovery, but I found a huge shot of motivation for myself. Never before in my life had I been able to transform my body in three days, now that we were sweating buckets, only eating when we felt like we had to, and walking 15-20 miles a day with a 30% of our body weight on our back, our bodies were becoming machines. My legs weren't screaming at me in the mornings as bad as they usually do, and Joey and I were in a good mood.

Real Food

We dismantled our tent in record time and hiked through the mist two miles back to the town we had driven through the night before. We walked around looking for a grocery store hoping to restock our food, but could only find a gas station. I was having trouble migrating away from the food court back at the University of Montana that was commonly filled with an array of deserts and food from around the world. Since we had started this trip, my king's diet had been replaced with pop-tarts, granola bars, and the occasional can of beans which Joey was fine with. I couldn't take it though, we needed fruit and something warm that was going to hold us over for a day. I could tell Joey did not want to sit down at a restaurant to eat. Luckily, I'm very persuasive and I don't think he cared enough to argue with me about it.

We walked into a little mom and pop diner with the shortest breakfast menu I have ever seen. Eggs and bacon, eggs and sausage, biscuits and gravy. *Well, alright,* I thought. *I guess it's biscuits and gravy.* We found a table in the corner with a place to put our bags. I left Joey to hang out and went to the front counter to find our food. I was greeted by a heavy-set woman with a jolly smile on her face. She had a whole hearted laugh she used when I told her I'd been living off granola bars for the past three days and said, "We'll fix that ASAP, sweetheart, what can I get you?" I ordered Joey and I each a full order of biscuits and gravy then we went back to my seat giddy with excitement. Joey looked up at me from under the brim of his hat and said,

Joey: You must be pretty excited about those biscuits, huh, pal?

Me: Dude, Southern cooking! There are whole TV shows just based on how good the food is down here! When was the last time you saw 'Idaho's next top Chef' on screen?

Joey: Alright alright... I'll give it a chance.

While Joey and I were bantering away, we had an older couple come up and ask us why on earth we were dressed like we were. With a smile, we shared our story again with new details we had collected from the previous day on the road. Soon, the whole restaurant was fighting to get a word in with us, but then I saw it. The heavenly gravy smell filled my nostrils as I watched our Southern chef walk out from behind the counter carrying two full plates of biscuits and gravy with coleslaw. Conversation could no longer be tolerated, I had to fill my stomach without interruption.

Joey was laughing as I excused myself from the conversation, assuring Joey and our new friends I would be back in a moment as soon as I destroyed these biscuits. I dug in and discovered the reason for obesity being so common in the South. All you need to start cooking like a Southerner is just stuff an absurd amount of fat and grease into any normal dish and be sure to include the least amount of vegetables possible. Not that I was complaining, there was an awesome flavor to them because I was so hungry, but I could imagine any person that wasn't walking 15 miles a day would be sure to die from a heart attack by 50 if they were eating those biscuits on a daily basis. I cleaned my plate and realized Joey had only cleared of half of his. "Are you gonna finish that?" I asked him. I didn't even realize I was already reaching across the table for the food he ordered.

With a look of amusement, he said, "Nah, Jake... eat up."

I downed the last half of Joey's meal and exclaimed, "Ah, yes! That was delicious, man, wanna try the restaurant across the street now?" For a second, Joey looked like his head was going to explode, I quickly caught myself, "Relax, man! I'm joking...geez." Poor Joey, I was so mean to him.

The Savannah River

We set off that morning with a smile on our face and full stomachs. Athens was 74 miles from the restaurant in McCormick, SC. We set a fast pace knowing that we couldn't expect to make it almost 90 miles like we did yesterday or be treated as well. If we could cover 30 miles a day, we might be able to make it to Athens the next night.

The sun had come out in force today, enough to drench us both in sweat before the end of our first mile. I looked over at Joey, "This is gonna be a long day, my friend."

Joey grimaced, "We'll be fine." Mornings were always the hardest parts of the day for us. Every time we would lay down to sleep, our bodies would try to heal. Well, no matter how long we slept, it was never enough time for our legs to recover or for the holes in our feet to heal. Because of that, the first couple miles of every day consisted of reopening the wounds on our feet and yes, it sucked. I'm pretty sure the nerve endings on both of our feet are now dead because of that first week.

So far, the South had not impressed me. We were in a beautiful environment, but the human sprawl had proven too much for me to overlook until we walked across the Savannah River. The bridge was built in between two hills on either side of the river, which gave us the highest vantage point for miles around. There was also a high concentration of parks and protected areas along the side of that bridge, ensuring that there wasn't a house in sight. The sun shining off the water in that moment melted my heart, that was the first time Joey or I had seen what I would consider a real piece of nature, and it is one of the few places I would ever go back to in the South.

Fear had been a huge part of our trip so far, people were so scared of us just because we were hitchhikers, we despised that. We could guarantee making anyone's day if they would pick us up but that didn't matter to them. They weren't going to give us the chance because of how homely we looked. Walking over that bridge, we were so taken back by the beauty that we stopped right there and stripped down to our shorts to show ourselves that fear may control most other people's actions, it would not control us, not today.

Joey told me I could go first so he could watch the bags. I looked over the side of the 35ft+ bridge and smiled, that was gonna be a hell of a jump. In full view of the passing cars, I climbed the railing on the side of the bridge. Taking a second to let out any nervousness I had run its course, I saluted my brother and leaped off the bridge backwards, arcing myself into a huge backflip. I counted 1...2...3 before I finally made contact with the water. I looked up and saw Joey cheering me on. *What a perfect guy to travel with,* I thought. I made the long swim back to land so I could walk back across the bridge and watch Joey repeat the process. I got such a kick watching Joey jump off the bridge after me, it felt like we were back in simpler times when I used to jump off my hometown bridge during the summers in Idaho.

Wait Up, Joey

We had used up a lot of time on our display against fear, we needed to get moving again. Redressing on the side of the road was entertainment in itself, watching the driver's eyes pop out of their head when they would see two cut hitchhikers on the side of the road never got old. I wiggled back into my shirt and looked over at Joey walking back across the bridge to see the funniest thing I had seen on the trip so far. Joey had the most perfect sunburn line I had ever seen down the middle of his chest. While I had been buttoning up my shirt throughout the trip, Joey left his open and now had a runway that started about nipple wide on his chest, getting progressively narrower and ended at his pant line.

I couldn't help myself, I cracked jokes about his chest for the next five minutes while he got dressed. He laughed me off, he knew as well as I did it was all in good fun (side note - Joey kept that burn down his chest for the rest of the trip!). I have never seen tanliness look better on anyone. I stopped cracking jokes, we had business to attend to, like walking 70 more miles. It was too late for me though, if I hadn't believed in karma before I made fun of Joey's tan-line, I was most certainly a believer about 15 minutes later when I said the words I never wanted to say on the trip.

Crossing that huge bridge didn't seem like much of an accomplishment until we saw the fabled sign we had dreamed about for the past couple nights.

Welcome to Georgia

"YEEEEESSSSSSSSS!" I screamed without giving Joey any warning and forcing him to cover his ears, not doing any favors to my already poor karma counter. We had made it! The first state line, even though it is just a line drawn in the sand by somebody a couple hundred years ago, it was still a victory and we had to take those wherever we could get them.

Working our way up a small uphill slope on the other side of the bridge seemed like every other uphill we had gone up on this trip. Nothing special other than it would be the place that the first complaint on the trip would be said. My foot was on fire again. Slapping my feet on the water when I had jumped off the bridge had opened up every old cut and blister on my feet. I was trying so hard not to say anything. Joey had passed me, keeping the pace I had set just a moment earlier and now could not keep up with. That went on for about five minutes before I finally said it, the dreaded words, "Wait up, Joey."

I couldn't decide what hurt worse, my foot or finally saying that which should not be said. Joey slowed down but even the slower pace was too much for me, I had to ask him to sit down against my pride. I took off my shoes and saw a blood stain from my glass cut on day one. Even if we took a day off to rest, that kind of cut was not going to heal in a day. That left us with but one option, keep walking. I bit my tongue and told Joey I was ready. He stood up and set a slow pace for me, looking behind him occasionally to make sure I was alright, which I hated that he had to do. *I am holding us back!* I thought to myself. I went total Focus Jake and repeated these words in my head, *If I can just last 20 minutes, the pain will go away. If I can just go 20 minutes, the pain will go away.* I recited my mantra and limped well through the next hour, never quite able to keep up with Joey. He was too good a friend to say anything, but it hurt me knowing I was holding my partner back.

I was hoping for a gas station so I could grab something to cover the cut in my foot. My prayers were answered, the next bend we saw exactly what I asked for. The funny party was that I had switched to my tennis shoes earlier to hopefully make my feet a bit more comfortable and now the blood from my cut had dried onto my sock. There was no way I was taking that shoe off now, looks like I was just gonna have to tough it out. Joey came back out of the gas station with a sort of acceptance on his face, before he could sit down with me, I stood up and said, "Let's get it, man."

There is only one kind of person in the world that doesn't feel pain; a dead person. I comforted myself with that thought while limping on my foot. Joey wasn't doing too hot anymore either, the back of his calves were cooked. Every step he took, the skin would stretch and cause what I could only imagine was the best feeling in the world from how he had shortened his steps so as to not flex his ankles. There really was a fair amount of comedy in our situation. My peg leg walk and Joey's penguin walk had to look absolutely comical from the perspective of a car driving past us. That was probably why we didn't have anybody swerve at us for those four miles of limping.

Trouble on the Horizon

We took cover under a huge billboard that was big enough to give both Joey and I a little shade. I learned something about Joey that day, when I'm in pain, I go deathly quiet. When Joey is in pain, he gets mad and unleashes a torrent of swear words that are so hard not to laugh at. Walking up the last hill, the only thing I had heard come out of Joey's mouth was, "Shit, fuck, fuck, fuck, shit, fuck, shit, shit, dammit etc." Then after we reached our break spot, he would whip out his e-cigarette and suck on it viciously, occasionally cursing at a driver that would pass by. It was so freaking funny, but I knew how mad he was so I would try to keep my mouth shut.

Joey and I are similar in the ways we think about the world, but our attitudes and actions towards it are often so different. I didn't like Joey cursing at the drivers, it was a curse and a blessing for me. On one hand was almost therapeutic in some ways because he would say everything I was trying not to think. On the other hand, even though Joey really didn't mean most of what he was saying, it was hard for it to not influence our mood on the really hard days. The thing that I hated most about hitchhiking with Joey, however, was that damn e-cigarette that he would whip out.

Looking back, I can see that my shutdown during those first couple of days could have made Joey turn more to his e-cigarette for a release because he no longer had me to talk to. While hitchhiking, I didn't even consider that possibility though. All I knew is that I would sit down after a really hard couple of miles, feeling proud of both of us until that sickly-sweet scent would hit my nose and I would look over to Joey, sitting upwind of me, sucking on his cigarette. That wasn't even the worst for me though. If I paid attention, I could time when Joey would start to get mad by how long it had been since he smoked. His mood would change in the matter of a couple minutes after inhaling on that thing. Very different from me, who never has, and never will smoke anything other than a salmon on a vacation to Alaska.

This was going to a bad place fast. I would get mad at Joey and bury my emotions with Focused Jake, and in turn, Focused Jake would piss Joey off so he would try to calm down by smoking and the cycle would continue. The fuses were lit, now when they would explode was the question. Luckily, the explosion was delayed by a kind and very religious man in a white truck. We hopped in the bed while he took us down some back roads so we wouldn't have to deal with losing our hats going 65 down the highway. After we stopped at the next town, he made a point to get out and ask us about our trip. We tried to keep our story short so as to not inconvenience our new friend, before he left though, he did the nicest thing for us. He shook both our hands and asked, "Would you mind if I prayed for you before you go?" Although neither Joey or I attend church or devote ourselves to any religion, this man was giving us something of his that was of the utmost importance. Graciously, we accepted. With a smile, he grabbed our hands and led us through a very long and detailed prayer, asking questions about our plans and incorporating them into his request of God along the way. I felt honored in the highest way and would have given that man a hug had I not been so sweaty. I often get a bad taste in my mouth from people trying to convert me to worship their gods, this man was different though. He didn't waste his breath trying to convert us, he was just kind and wished us the best in the only way he knew how.

Will Sing and Dance for Ride!

We spent a long time in that next town looking for cardboard and an orange. Literally ran up and down the street for an hour asking every single store if they had any cardboard or fruit, getting the same answer every time, 'No, but try (*insert store name here*).' Finally, a hardware store gave us a plasma screen TV sized box and showed me their orange produce in the back. Now fulfilled in my quest to find cardboard and some fruit, I could finally relax and carve some signs out of the cardboard.

These were not going to be our typical destination signs though, we had to make this some of our best material yet! How were we going to get the attention of all those cars that drove by us? We had to think for a couple minutes, so Joey turned on his phone and started singing some long black train by Josh Turner. "That's it!" I exclaimed. Joey was sounding like a country star in the making, so why not make his sign to read, 'Will Sing For Ride.' I was willing to take back up on this one and be his background dancer so my sign read, 'Will Dance For Ride.' Satisfied with our creativity, we set off into the Southern heat.

I don't know if you would consider it lucky, but we didn't have to deal with the heat for long. It turns out that there was a massive hurricane that had been following our path for the past couple of days and had finally caught us. Utter downpour commenced as Joey and I found our way to where the highway turned into what would have been a freeway back in Idaho. Two separate roads and three lanes each was apparently still a highway in Georgia.

O my god, I have never seen it rain as hard as it did for the week that this hurricane tagged along as a third member of our hitchhiking party. That freaking storm would drop a lake on top of us in the span of about 10-15 minutes before it would settle into a drizzle for a couple hours and repeat. Something I love about Joey and I's friendship though is that we love the rain, relish in it actually. Rain is just such a beautiful force of nature that cannot be looked over by the minor inconvenience of being wet. The only sad part was that I thought it would hurt our chances of someone letting us in their car. Against my best predictions and in favor of my greatest hopes, a young man pulled off the road in front of us with a smile on his face. First thing he said to us was, "Well, are you gonna sing?" We laughed, we had been walking so long in the rain, we forgot we even had signs on our backs. Joey broke out into some old country and I did a backflip for good measure. The kid laughed and let us in, we both hopped in the back this time and got the best news we had heard all day.

The Kid: Where are you guys headed?

Joey: Athens, as close as you can get us would be great.

The Kid: I'm actually headed back to Athens now. I can drop you off there.

Me: O my god, I love you.

That line might have come out at the wrong time because I looked down not two seconds later and saw that my feet were in a pile of panties. At first I thought the kid was just a player, but as the drive went on, it became very clear he was gay. Not that he wasn't a great guy that took great care of us, we paid his gas when he filled up and had a great talk. Buuuuut 'I love you' probably shouldn't have been the first words out of my mouth. He played some of his own personal music for us over the 30+ mile ride into the heart of Athens. We said an awkward goodbye as the kid lingered to talk to us while we stood in the rain. We listened out of gratitude but cut the conversation short when it got too long. We had to find a place to sleep where we would not be soaked by the rain.

The Storm Inside the Storm

Joey took point on finding us a place to sleep on his phone, apparently there was a park about five miles away. I don't know why, but I decided right then and there would be the right moment to argue with Joey. There was just little tension in the air for some reason or another even though we had just made it to our destination. As we walked, I shut down after trying to argue with Joey about where we should sleep. I didn't make it easy for him, but Joey convinced me to follow him into the city. In silence, we walked, I took the lead even though I didn't know where I was going, forcing Joey to shout commands from behind me on which way to turn. The tension became too much after the first three miles and Joey confronted me about it. I think part of the problem was that I am a much more emotional thinker than Joey, using more of how I feel to decide than logic like Joey was doing in our walk through Athens. No idea what we said, but I did tell him that I was just focusing, not mad at him. That is the moment where Joey coined the title 'Focused Jake' and let it be known, Focused Jake is a dick.

Both of us exhausted and drenched, we reached the park. Joey led me past basketball courts and across a long grassy field before we found hobo paradise with a plug in for our phones, water, and bathrooms only a short walk away, even a roof to protect us from the rain. The minor downside was that it was directly in front of the door to the security room for the park. There was a man we could see inside forcing us to wait a distance away from the door so we didn't want to seem like we were waiting to break in. It didn't take long for the man to wrap up his last duties and leave, allowing us to foster the start of our first hobo club. The bad news was that there was still unsolved tension between Joey and I. Everything has a silver lining though, tonight we were sleeping on concrete for the first time and in doing so, reached hobo level 2.

Getting out of Athens

I woke up feeling much better, something about sleeping on concrete straightens out all those knotted up parts of you. I looked over at Joey to check in on how we felt about each other, not much had changed. I didn't know how we were going to fix the rift that was coming between us, but there was no time to worry about it this morning. It was still early morning and we did not want to have to sleep in the city for another night. Not wanting to walk through the bad parts of town, we decided we would take a bus.

Joey took the role as navigator again to get us to the nearest bus station. Knowing that Joey's detail-oriented mind was going to get us there much faster than my big picture approach, I fell in line. It was in both of our best interest if I just listened to him today. We walked another five miles in moderate discomfort due to the rift that was growing between us. Most of the time, we could finish each other's thoughts and know what the other wanted before they said anything. That morning passed with a strange sense of unfamiliarity; not that we weren't still brothers that would finish this trip, but it was strange knowing that each of us were thinking individually instead of as one for the first time.

We made it to the bus stop well before 8 a.m., the busses were supposed to come in 15-minute intervals according to the sign next to the stop, but there were no busses to be seen. It was probably good for us, as it forced us to spend more time together to deal with our issues. Aside from my pain being somewhat less disabling than the day before, neither of us were in a great mood after that walk. We stayed mostly silent until 9 a.m. when we realized that we could have covered 3–4 miles in that hour we spent sitting on the bench.

Joey stood up and took point before I could say much and started walking faster than he had on the entire trip. *Maybe we just needed space from each other,* I thought to myself. I did the best I could to keep up with Joey's pace up the hill until I started to feel things cool off a bit between us. Walking across a huge bridge over some of the greenest water I have ever seen, I caught up to Joey and tried to bridge the gap by apologizing for being so focused the night before. Joey brought up the point that we knew we were going to hate each other a couple times on this trip, this was no surprise to be a little mad now that we had been hitchhiking for five days. It would pass, but before we resumed our silence, I made a mental note that Focused Jake was what started our problems in the first place. "Just tell me when I get too focused, alright, Joey?" I asked him.

He said, "Sure," in response but I knew I had to change if we were going to keep this partnership together across the country.

Around 10 a.m., a gentleman was kind enough to drive us the rest of the way out of Athens. As another former hitchhiker, he told us stories about his brothers and he used to ride the trains all over the country. Boasting that his brother had once gone from Seattle all the way to New York on the train. "What an adventure," I said to the man. He didn't stop with his stories of adventure all of the 15 or so miles he drove us. It made me think of the generations before the one that Joey and I find ourselves a part of. Where had that sense of adventure gone?

Where Did the Adventure Go?

Have you ever really listened to people's stories? From your grandparents to the homeless people on the side of the street, everyone has such a story to tell. What I've noticed especially after hitchhiking, is that the younger people get, the less exciting their stories become. Both Joey and I's parents tell the most fantastic stories of when they were kids. My dad used to hotwire the neighbor's boats while they were on vacation and take them for spins out on the lake with all of his friends. Joey's dad was a street renown brawler that people would seek out to see if they could beat him, though NOBODY could.

If you compare those very surface level stories with those of our generation? It doesn't get much more exciting than someone falling off their bike and breaking their arm, someone meeting a girlfriend of five years on Tinder, sometimes even the real bland stories of, 'When I was 13 *sobbing* my boyfriend broke up with me and *sobbing still* it just really hurt me, you know?' It is easy to see there has been something lost on our kids over the past 30–50 years. Joey and I are one of the few exceptions to that broad observation. We can now say that we hitchhiked across the country, had people with guns tell us to come outside with them (wonder what would have happened), and we starved in the woods for three days. None of that is embellished, it is just what happened and we've been able to turn it into a great story. You can turn meeting your girlfriend on Tinder into a great story, but the substance just isn't the same as the stories the generations before us have to tell.

The Death of Focused Jake

That man had put us in a good mood and dropped us off with a skip in our step, but it didn't stay that way. We knew that we couldn't count on another ride, but it didn't make the hours and hours of walking any easier. Another torrential downpour came and went, now the sun was beating down on us yet again. Joey's chest and calves were basically glowing from the amount of sun he got and his refusal to button up his shirt or zip on his pant legs just gave me something to be mad about. I almost said something, but caught myself right before I blurted out how dumb I thought he was for not covering up. I was not going to keep feeding into this negative loop that was pushing Joey and I further away from each other. I needed some space to let everything out. "I'll be at the store up the road," I told Joey. He could spend a little extra time at the gas station we stopped at while I cooled off.

Breathing deeply from running with my bag on my back, I burst inside the grocery store and bought Joey and I our first taste of real Georgia peaches. This would be a good peace offering to bridge the gap between us. Joey walked around the corner after a couple of minutes and I gave him my gifts. It was nice, but I still needed to step up my game if I was to end the terror of Focused Jake. The peaches softened our mood and allowed us to have the first halfway decent talk in hours, but silence commenced as soon as we hit the road again. We took to opposite sides of the road almost instinctively, the 20 feet of asphalt between us serving as a void between worlds. I opened my mouth several times to say something, but the tension would steal the words out of my mouth each time.

Taking a deep breath, I broke the silence and yelled through the wall of cars between us, "Hey! You wanna race up this hill?" Joey looked at me for a moment, maybe surprised? Then his mouth stretched into the classic Joey grin that had been missing from our trip for the past day. "1...2...3!" We took off up the hill, the cars that had been flying by us slowed down to watch the hobo race. I snuck a quick look at Joey, *Damn, he's fast.* I spurred my legs a little harder and ran up the hill with all I had left. Finally reaching the top, I hunched over on my knees wheezing. Looking up, I saw Joey doing the same thing, "Call it a draw?!" I gasped. Joey's chest bounced up and down, his laughter stealing a breath from him. He caught my gaze for a moment and flipped me off, sending me into a fit of laughter. Consider the rift closed.

Hooters

Joey and I made camp that night in a beautiful stretch of woods that gave us a taste of what the Southern forests were like before our ancestors sailed across the ocean. We woke up on day six with a different feeling in the air. That is the first day I can say I truly felt like a hitchhiker. Both of us knew what to expect for the most part now, so we accepted it. Of course, the day you think you know it all life throws another curveball at you. Today would be the day we make it to Mcdonough, Georgia and step into a whole new way of living.

Our morning passed in relative comfort, some of the same nice people that had picked us up the day before picked us up again and gave us a ride for a couple more miles. Another short ride from a former army medic left us just a couple blocks from temptation. A Hooters sign let us know that we were just a few blocks away from pretty waitresses in tight tank tops. We looked at each other and gulped.

Jake: There are more important things to do than flirt with girls, right?

Joey: O, definitely. We still have such a long ways to walk today, we don't want to waste any time…

Jake: Great… It's decided then.

We looked at each other, we had plenty to eat already, but it's Hooters, right? I'll admit I was definitely wanting an excuse to talk to anybody our age, especially scantily clad college girls.

Ten minutes later, we busted through the doors of Hooters as their first customers of the day. As usual, every staff member's eyes looked us up and down from several times when we walked in. Their looks told us that we were definitely the first Indiana Jones looking dudes walking into a Hooters of all places. I remember Joey having the biggest grin on his face at the sight of seeing our waitress. She was the first beautiful Southern girl we had seen. Coming from Idaho, we don't see a lot of other people other than white Americans. This girl was gorgeous in a way we had never seen, she was black, just over five feet tall, and had the most adorable smile that melted our hearts with the first word out of her mouth. Joey and I were in love.

We sat there and shot dreamy eyes at our angel coming and going from the kitchen. It took me a minute, but I snapped out of cupid's trance when she introduced herself. I found my words again and talked to her about the only thing we could, hitchhiking. There was no way to talk about without sounding like we were a couple of cocky kids off the street trying to come up with a great story. I could see her cogs turning in her head, analyzing us to see if we were legit or not. In her moment of confusion is right when Joey snatched his opportunity to dive in. All I remember is him saying is, "You have a beautiful name." From there, the chase had begun.

"Oh, uh, thank you," she said and turned to leave. She had just thrown out a road blockade to keep Joey from advancing.

What she didn't know is that Joey has seen a blockade or two in his time and vaulted it with the grace of a gazelle by saying, "Before you go, what would you recommend we get here? We're leaning towards the dry rub wings, but we know Southern barbecue has a reputation." His gazelle leap had paid off, she turned right around and gave us just about the most fun we had so far in the day. She gave us a recommendation and laughed with us for a bit, but she had to get back to work. Joey leaned back in his chair and smiled at me, "She'll be back," he said.

My god he was right, our wings ended up taking a little long so our angel of a waitress walked back out with a free plate of fries (tip for all you ladies out there; food = all the men you could ever want). Wings and fries ended up being a perfect energy shot for us. Joey bought for both of us and continued to try his luck with the waitress. We learned a lot about her, a college student in Athens living the sorority life. I asked if they ever got hazed in the South and she gave me the kind of college girl grin that makes you want to move across the country just so you can see her again. "That's a sorority secret," she winked at me. I laughed, not just a pretty face, she had a personality to go with it? That is something that you don't see all the time and I appreciated it deeply. Despite my protest, Joey would later leave our Instagram page on our bill. We never heard from the girl again and part of me likes it that way. One of a man's greatest inspirations he can have is a woman to fight for. That Hooters waitress had given us that for the time being.

Drug Dealers are People Too

We had another big walk on our hands after that. The South did its thing and dropped another rainstorm on us. It was like being thrown in a never-ending cycle between a washing machine and the dryer. The rain would dump on us and was away all of the sweat and grime that would accumulate on us, then the sun would bake us to a crisp. We never really got dry because of the 80–90% humidity that made us feel like we were walking on the bottom of a lake, but that didn't stop the sun from burning every inch of skin left exposed.

By day six, we were used to such treatment and Joey was getting restless. He had found something on the maps looking at his phone, some sort of old monastery looking thing seven miles up a county road had caught his attention. I had no desire to go at all, but Joey is persistent, we spent about five minutes arguing before I gave in on one condition. If someone stops to give us a ride, we were heading to McDonough. He reluctantly agreed to my terms while I hoped that we wouldn't get stuck anywhere without food.

We walked for a couple more hours until we were about a mile away from the turnoff to Joey's monastery. I had accepted the fact that we were going to it, there was nothing I could do. I was noticing Joey pick up his pace the closer we got, he was so excited. I had accepted my Fate, but the moment that we were about to start that seven-mile trek, a big truck pulled off of the road in front of us. That moment was the only time I saw Joey upset that someone was stopping to pick us up. I watched his excitement completely deflate as I told the driver we were headed to McDonough. With a warm smile, he welcomed us into his vehicle. I hopped into the front seat, leaving Joey plenty of room to sit in the back but he was not about to get in the same car as me. He looked right through me and told our driver he was going to sit in the bed. *Huh,* I thought, *that is going to make for an interesting talk later.*

That left me in the front to talk to our new friend and former meth and heroin dealer who gave us one of my favorite stories on our trip. I have no idea what I said to make that man like me so much but before I knew it, he was insisting that we come see his new house that he was building. I was skeptical for sure, but having someone else that both of us could talk to would be good for Joey and I. Making the executive decision for both of us, I accepted the man's offer and watched as a huge smile stretched across his face. With enthusiasm, he assured me of how good of a choice I had just made. *Okay, the guy is a little crazy, but I can dig it,* I thought to myself.

It wasn't long before we pulled up to his house under construction. "Oh, you're having it built for you?"

He grinned again, "Nah, man, this is all me." I could see now why he wanted us to come back so bad, his house was turning into the most beautiful little home all by his own hand. He had told us he was a construction worker, but now seeing his home, it was clear this man had a real skill. The house had a personality about it that most new homes are missing. His specific taste showed through in mismatched lights and hand-cut boards that made up his porch; none of it was traditional, but it felt like it worked.

The car came to a stop and Joey jumped out of the back, just the sound of his feet hitting the gravel told me he was still mad. The man we were with must have noticed as well because he invited us into his trailer for a beer before we took a tour of his growing home. Leading us inside, the man pulled out a couple of beers from the fridge and offered us a drink. Joey graciously accepted the

Heineken he was offered while I elected for water. You could feel the positive energy in that little fifth wheel as the man gave us our drinks and sat down to tell us some stories.

It turns out he was on the FBI's top 100 most wanted list for three years previous to life he has now, not what you would expect from such a kind man. He had spent his 20s and early 30s building a street cred for himself as one of the best and most uncatchable drug dealers in the South, quite a reputation to have. He had avoided getting any dirt on his hands for years until one day he received a call that he credits as, "The best change that ever happened to me." The call was from the local sheriff's department to let him know that he was wanted for drug smuggling among other crimes and was to report into custody immediately. He said he hung up the phone and took a couple of deep breaths. He always knew this day would come, but now he had the decision of what to do about it. Should he turn himself in and take the lesser sentencing? Or should he run and keep making a ton off of his drug dealing business? After a couple minutes of thought, he chose the latter option. In his own words, "It was on man."

He snapped his phone in half, abandoned his car, dyed his hair, and stole himself another ride all before the night ended. If he was going down, it was going to be with a fight. For the next 18 months, he lived on the run, narrowly avoiding the police on multiple occasions. The guy was a total pro criminal that the police could not catch, that reputation earned him a spot on the FBI's most wanted list. That only fueled his want to avoid the police though, being on the FBI's most wanted list was a badge of honor he could wear among the criminal underworld. What he didn't count on was that his fellow criminals were not as good at avoiding the law as he was. Hanging out in a meth house on the outskirts of Atlanta one day, he said the FBI surrounded the building, tipped off by an anonymous call for one of his buddies that had not been as discrete. He ran through the house to his hiding spot he had made sure to have just in case something like this happened, but stopped cold in his tracks when he heard the words that would bring his career as a drug dealer to an end. A voice came on one of the speakers outside, "This is the FBI, we have permission to use deadly force if you do not come out with your hands up!"

Halfway to his hiding spot, our drug dealing friend stood stone still in the hallway. He wasn't scared of death; it was part of the thrill for him in his chosen career. He knew something that the law enforcement outside didn't, there were two babies in the house. Looking over his shoulder, he could see one of them crying in it's drugged out mother's arms. He debated for a moment whether or not he could live with the guilt in his mind if that baby was killed so he could continue his life of hiding. "In the end, there was no decision," he told us, "I had to give myself up or risk that kid's life." Honorably, he took the longest walk of his life back down the hallway past the crying baby to the front door. He took a deep breath and tried to talk himself out of what he was about to do, but it was no use. Scared out of his mind, but knowing he was doing the right thing, he opened the door and walked out with his hands up.

Joey and I were entranced by this man's story. He had a rough past that would lead many to think he is a bad person, but you couldn't be more wrong. That man held himself by a code of honor that most law-abiding citizens do not. He took a moment to look at the floor and relive the experience in his mind. He laughed, "I don't want to talk y'alls ear off though! Let me show you what I've been working on." With that, he took us on the grand tour of his almost completed home that represented two years of work and dedication on his part. The house was very plain, but it remains the most spectacular house tour I have ever had.

While in prison, our friend managed to get involved with some sort of construction work that allowed him to go to the outside world once a week to do free labor. He said, "As soon as I started, I knew that construction was what I had been missing my whole life." He spent every hour he could on the construction crew building new houses. Years of doing that work for free gave him a set of skills that he uses in his own personal construction jobs now that he is a free man. He told us everything about the house, what troubles he had and what had been the most fun for him. He was not bragging, simply proud that he had turned his life around so effectively. While most men would go back to dealing drugs (especially someone that was so good at it), he had found a better career that he enjoyed more than the high of being on the run. Laughing again, he led us back out to his truck and told us he would take us anywhere we wanted. Joey sat in the cab with us this time and told the man to drop us off at the first gas station in Mcdonough.

The story you just heard is one that has never been told. I have never told anybody because the power behind it seems to be too much to incorporate into any normal face to face interaction. Whenever I feel hopeless or scared of what the future holds, I think back to that man and his story of losing absolutely everything. In jail, he truly thought that he would never have anything again. To come from that and be building your own house just a couple of years out of prison is the ultimate example that it is never too late to start over and become everything you never thought you could be.

Racism

A town that is basically just a suburb of Atlanta, Mcdonough is the spitting image of the perfect suburban life. Beautiful trees and flowers in every yard, nice winding roads never going above a 25-mph speed limit. It seemed a nice place to grow up. Our former drug dealing friend ended up driving us all the way to Joey's aunt's neighborhood. We didn't have to walk far before we found the correct address and stood in the driveway a moment like we had at the Metzger's house. "Alright, Joey, your turn to knock, pal."

"Yeah, yeah, whatever, Jake," he retorted with the classic Joey grin. Before we even got the chance to walk up the driveway, we saw a group of people come out of the garage. They were only about 15 feet away and had not seen us yet, we didn't want to spook them so we just stood there waiting. Eventually, the mom of the group caught us out of the corner of her eye and screamed.

Joey and I ducked into athletic positions so we could run quickly in case we were walking into the wrong house. Our suspicions were silenced when the lady screamed again, "Is that you, Joey!" A moment later, Joey was wrapped in hugs from his long-lost family, it was absolutely adorable. There was Joey's aunt, uncle, and his two cousins; a 20 year-old-man and a 10-year-old girl. Apparently, the last time Joey had seen this family was back when he was a toddler. They showed us the picture of him and his cousin in little suits back when they were both little kids. Everything felt so warm and inviting walking into that house in the suburbs. They took us inside and sat us down, keen to take care of us in every way they could. We did our best not to be a hassle, but they insisted on feeding us. 30 minutes later, we were both showered and eating delicious ice cream. Such a peaceful evening didn't seem like it could be spoiled by even the worst turn of events. Being wrong was starting to become something I was pretty good at though. Who knew that our time with that family would turn so sour?

Joey's aunt decided that she was going to run to Jimmy John's and get us some sandwiches. We couldn't say no, that sounded amazing to have some more good food. She left in a hurry to Jimmy John's, the son left in his truck to take his girlfriend to a Kenny Chesney concert, and the little girl left to a friend's house. It had gone from hustle and bustle to just us and Joey's uncle in the house. First impression of the man was of a gruff veteran that would still be out there in war if he had the option. The poor man was suffering from kidney stones, which left him confined to the big chair in the middle of the room to watch the TV for the majority of our time with him. The more we talked, the more I ended up liking him, this family seemed pretty great. I was sure it was going to be a lovely evening until we got that phone call.

Joey's uncle picked up the phone to find his wife (Joey's aunt) on the other side of the phone. We watched as he got more and more panicked while talking, asking over and over again, "Are you okay? No, are you okay?" With curiosity, we watched until he hung up the phone and told us that Joey's aunt had been in a car crash on the way to get sandwiches for us. Of course, we asked if she was okay and he told us she would be as long as he could get to her. It was admirable to listen to him talk about his wife in that way, it seemed to be real love. The problem was that his son had taken one car to Kenny Chesney, and the aunt was in the other, which left us with no cars at the house.

For the next couple minutes, we sat in silence and watched Joey's uncle call every single one of his friends until one finally said they would come get him. He was so troubled at not being able to do anything but wait that he distracted himself by ordering Joey and I to the kitchen so he could teach us how to use a microwave to warm up frozen fish sticks for 20 minutes. We knew he was in pain so we didn't say anything, we just let him say whatever he wanted to say. As time went on, his worry turned into anger at whoever had hit his wife. It was understandable to be mad, but what came out of his mouth next was not.

Side note -

Note that this next part is extremely hard to write. Even though these people that we stayed with were related to Joey, I want to clearly state that Joey and I are not racist and never have been. We use the language we do in this book to portray the shock that we felt when encountering these situations. These words are not meant to hurt or negatively alter anyone's opinions towards any race of people. This is only to give you a true insight into what we experienced firsthand on our journey.

"I KNOW SOME BLACK BITCH HIT MY WIFE!" he thundered. Joey and I were totally caught off guard. He proceeded to tell us about the only way someone could have hit his wife is if they were black because only 'niggers' would be stupid enough to pull out at a red light when it is not their turn.

It was as if our words had been stolen from us, my jaw dropped listening to the tirade of racist insults that now filled the house. Our only hope of stopping was for his friends to show up and take him away soon. Like a gift from the heavens, car tires on gravel cut the uncle off mid-sentence, his friends were finally here. Already filled with rage, he tried to run out his garage door but tripped and fell a couple feet down to the hard concrete. His knee came down hard enough to hear 'BOOM' echo through the house. Joey did the best he could to help him limp the rest of the way to his friend's car while he cursed the whole way there, not stopping even as Joey closed the car door behind him.

I had already left the scene; I couldn't take anymore of the hateful speech that was coming out of his mouth. It wasn't even necessarily that he was talking about racist ideologies that Joey and I both despise; it was the hate that filled him. If he didn't have black people to complain about, that hateful energy would just find something else that he could freak out about.

Joey came back inside to find me sitting on the couch thinking. I was doing my best to take the hate out of his uncle's words and really understand what inspired him to say that. After talking it over with Joey, I realized that he was only hateful because of the same thing we had been running into the entire trip: FEAR. He spewed out racism because his wife was hurt somewhere and even though he loved her more than anything in the world, he couldn't do anything to help her. Therefore, all that loving energy transferred itself to the first other thing it could find, a lifetime of being told that black people are inferior to him.

I'll admit, I didn't want to deal with it. I thought if we went on pretending that the tirade of racism didn't just happen, we could try to enjoy our time with people who had been very good to us. We did exactly that, we let it go, and in doing so, let that formation of a new mindset start to take place under that happy and joyous one that we were trying to get back to.

Joey and I were alone in the house for hours before the aunt and uncle finally walked back through the door. Joey's aunt was fine other than a couple of bruises, that didn't stop us from leaping into action and attending to every want and need she could have for the rest of the evening. We had mostly forgotten about the earlier drama that took place and did our best to have fun with Joey's family. We truly enjoyed the aunt and her humor in the unfortunate situation.

More Racism

As our first night in McDonough came to a close, Joey's uncle took over the conversation to tell us, "There's something I've been wanting to tell you boys since you got here." We turned our attention to the head of the house and hushed ourselves. He took a moment to fill his lip with a clump of tobacco and spit in a cup before he spoke. Pointing his finger still covered in tobacco at us, he said, "When I heard you two were gonna do this a month ago, I thought y'all had a death wish. I know you boys are friendly and good at making friends, but there are people down here in the South that don't wanna be your friend, that will seek you out even if you ain't done nothing to um, not just to tell, but show that they have a problem with you."

Joey was the one that held the most authority with these people. I was only Joey's friend and my opinion didn't hold much value within these walls. Because of that, Joey did most of our talking for us. My brother replied with intensity, "That's what we want to show people though! Most people are good, we are out here to prove that humankind is not a lost cause and that we still can help each other." I could hear the passion in his voice, it is so inspiring to hear somebody talk about something that they firmly believe in, especially Joey. His uncle must have heard the same intensity and passion in Joey's voice as I did because he withdrew his pointing finger and recoiled in his chair to break eye contact and avoid the flicker of fire that lights in Joey's eyes when he talks about something close to his heart.

I was very interested to see how this would turn out. Joey is not one to let opinions pass without first questioning them, now he was questioning the dominant male in the situation. There was a shift in the air that you could feel, Joey now had the floor and had taken position as king of the house for just a moment, then he realized who he was challenging and backed down out of respect. There were a few seconds of silence that hung in the air; who was in control now? Joey's uncle leaned forward in his chair, maybe the uncle felt he had to make up for the brief shift of power between him and Joey, because he kept us up for well over an hour after that assuring us that we were wrong and that most people are out to get you. His words: "There are a lot of people looking to take advantage of a couple of boys that don't know what they're doing…especially black people." The racism was not a one-time thing as we had hoped, he was trying to drive his beliefs into us by repeating himself over and over. "Those black people are out to get cha," he kept saying.

Logically, we knew none of this was true, but we couldn't say anything because this family had treated us so well. Even after the earlier car crash, which would have been plenty of an excuse to tell us that we could not stay with them tonight, they had gone upstairs and set up our own room for us and bought us food. So many kind deeds seemed to be a bribe for the uncle to now say whatever he wanted and have everybody agree with him. I knew neither Joey or I were going to take any of this to heart, but it was not the first time we had heard it. Multiple people that had picked us up had dropped little racist tidbits here and there that we would try to ignore. When you absorb opinions for so long with no ability to voice your own, the subconscious mind starts to shift towards that belief because it's the only information it's being fed.

Joey and I went to bed that night without saying much, that conversation downstairs had been a roller coaster of emotions that I couldn't begin to voice. We had thought we would leave tomorrow, but now that both Joey's aunt and uncle were slightly incapacitated (uncle with a kidney stone and aunt from the car crash), we felt obligated to stay and help them out for another day. I felt trapped between a rock and a hard place. I wanted to leave but it was only right to stay and help Joey's family. The upside was that there is nothing like a shared hardship to bring brothers back together.

The next morning, we woke up early to get breakfast with Joey's aunt in a great restaurant downtown. We were happy to accompany them until we saw where we were going. We parked close to a grassy area with some sort of memorial in the center of a flower garden. Enjoying the freedom to run around without the 40+ lbs. on my back, I vaulted the short brick wall surrounding the park and ran into the flower garden. I was expecting the memorial to be for Vietnam, but what I saw made me question if Joey and I had made a mistake by coming to the South. It read something along the lines of,

'Let This Be a Tribute to the Brave Young Men of the Confederacy That Fought the Union to Protect Their Southern Values.'

My blood went cold, I remember looking around to make sure that I had not somehow stumbled onto KKK territory. I felt relieved when I saw a black couple walking down the street, but then I took a moment to think about what that memorial meant and my relief turned into anger. *Protect their Southern values?* I thought to myself. *The Southern values that supported slavery? The Southern values that would not have let the black couple that just walked down the street to walk into the same restaurant we were going to? The Southern values that would protect organizations like the Klu Klux Klan after the Civil War?* This memorial was a reference to the Civil War from 1861–1865. While the war was not fought directly over rights to slavery, slavery was one of the 'southern values' that the Confederacy fought against the Union for. This memorial was honoring old and twisted thoughts that should have been long forgotten by society, yet here this memorial stood in the center of town as a cruel reminder of our past mistakes.

I was blown away, how was something like this legal? How is there not an uproar to tear it down by everybody that walks by? I looked around that little town square expecting to see an angry mob ready to trample me, but I saw nothing. From that moment on, I looked at the entirety of the South in a different light. Although the hateful words are no longer used in public like they used to be. The values that racism is built on are still there, allowing those hateful words and ideals to still very much exist, especially behind closed doors.

I walked into that restaurant with Joey's family on edge, before I came to the South, I thought that racism was a thing of the past with a couple of small and isolated pockets that still existed. Now that I had seen a Confederate memorial in the middle of town, I didn't know what else to expect. I might as well see a werewolf run across the street and a cow jump over the moon. Point being, none of this was supposed to be real, but I was living in it.

I felt sick and breakfast no longer sounded appealing to me. It was almost as if any act of consumerism would continue to support the terrifying mindset of racism. My stomach growled and assured me that my body needed food though. We had not been getting enough calories for our constant walking and had to make up for the deficit at some point. I choked down some grits and ate the leftovers on the little cousin's plate just to keep up appearances. Joey was eyeing me suspiciously; he could read me like a book and knew something was wrong. On the way back out to the car, I told him to walk with me so I could show him the inscription on the Civil War memorial. Pointing it out, I said, "I didn't know this stuff was real anymore, Joey." He went quiet for a moment as he bent down to read the memorial.

Standing back up Joey told me, "It is very different down here… We can leave tomorrow and try to get further north, but let's try to enjoy the beauty as much as we can."

I was initially taken back by Joey's approach, unable to see how he was able to take in the fact that racism was not only real, but still prevalent, in stride. However, the more I thought about it, the more I realized he was right. I was letting an externality dictate my mood and actions towards my environment when I didn't need to. Neither of us were about to agree with the environment we found ourselves in, but we couldn't give into it by responding to the brutal reality of racism with more hate and fear. Joey was dropping wisdom on me like it was hot. We spoke in hushed tones for a couple of seconds before we were called back to the car by Joey's aunt. That was one of the many, many times that Joey saved me on that trip. He even got me to sing some Luke Bryan karaoke on the way home, but the trial of racism was not over.

"Daddy, can I go to my friend's birthday party today?"

Later that day, we met up with Joey's uncle to go get their crashed car out of the impound lot. It was a Sunday, and the appointment we ended up making to come get the car just happened to be right when most church services were ending. Being in the center of the Bible Belt, that meant A TON of traffic. We were actually quite entertained, everybody in sight was coming from church in their Sunday best. It was really cool to see a whole town in suits and dresses; I was loving it. Joey's uncle, on the other hand, was furious.

The whole car could tell he was bottling it up, it was why Joey I started hosting our own comedy show in the backseat to alleviate some of the tension. It worked until we passed a predominantly black church with so many people coming out that traffic came to a complete stop. It was the final straw for the uncle. It started off without any racist remarks, "Every time we pass this damn church we have to get stopped. Why do they have the audacity to think that they have more right to the road than us?" Sitting in the back seat, it was easy to see where this rant was going. We tried to stop him by joking about how malicious the impound lot was to schedule us on this exact day, but it was too late. I won't quote what he said, but let's just say there was a long string of hard R 'N-words' that followed soon after.

We went quiet, there was no use in trying to stop him, anything we said would only fuel the fire. Joey's aunt let it continue for a long time before she tried to calm him down, probably more to keep his blood pressure from getting out of check than to stop the hateful speech. I looked to their daughter sitting with us in the backseat with us to gauge her reaction, keep in mind this girl was 10 years old. She didn't flinch throughout her dad's whole rage episode. *O my god,* I thought. This was not a one-time thing brought on by being in a stressful situation, the uncle was mad like this all the time.

We arrived at the impound lot and paid for the car without any problems. I was so relieved by the thought that Joey and I were going to get to leave sooner! It was just wishful thinking though, we still had a whole car ride back home that could go wrong. We made it almost all the way there without any problems until Joey's little cousin asked her dad, "Daddy, can I go to my friend's birthday party today?"

Without even looking at her, he responded with an icy cold voice, "Are there going to be any little black girls there?" The little girl didn't know how to answer that question. Before she had time to think, her dad repeated himself louder, "ARE THERE GOING TO BE ANY BLACK GIRLS THERE?" You could see in the little girl's eyes that she knew the stakes had just risen, she had to say something.

"Ummm...no daddy," she whispered.

Just then, we pulled into the driveway and I left the car as fast as I could. I was so angry that Joey's little cousin wasn't even getting a choice in racism, she was being taught hate from such a young age. If I stayed in that car for a second longer, I was at risk of saying or doing something I really would have regretted. I retreated to the upstairs room and did my best to center myself with some meditation. *This will only affect me if I allow it to,* I told myself. Joey joined me after a couple minutes and sat down on my left, "We are leaving tomorrow morning," he said. There was no room for argument in his voice. That last episode from the uncle was too much for either of us to handle.

Release

Joey's older cousin and his girlfriend drove us 40 miles around the bottom of Atlanta early the next morning. I was very grateful we didn't have to ride with Joey's uncle again, I felt set free from a prison and actually did enjoy Joey's older cousin and his girlfriend a lot. They were travelers that had driven to all the nicest beaches in the South. It was the first wholesome conversation we had since getting to McDonough two days ago.

We reached a mall parking lot and Joey assured them that we would be fine if they dropped us off there. They were so kind that they tried to give us cash and bury us in food before we got out, but Joey and I could not accept anything. We needed to finish this on our own. With that, they drove away, leaving us with emotional turmoil to process.

For the first time in two days, Joey and I were able to talk freely about what we had just experienced. I didn't feel fully able to express my feelings, however, there was just too much I wanted to forget for me to say anything about those last couple days, so we defaulted to small talk and walked. There was one nice family that stopped us and told us their grand plans for doing exactly what we were doing. "We are going to take our kids and hitchhike from New York to Seattle!" they exclaimed. That cheered us up a bit, we were back to inspiring people, and that felt incredible after the ordeal we had just been through. We thought we were getting better; on an upswing you know? Then there was that kid…

We were entering the rough neighborhood, you could tell by the bars on the windows, the trash all over the ground, and the people's refusal to look at us in the eyes. It was understandable for us to be on edge in such a place, but it was not okay for us to become part of it. We rounded a corner and saw a black man sitting on a small brick wall (one of the biggest changes I had noticed the first day away from Joey's extended family was that I was first noticing people's skin color before talking to them now, and I hated that about myself). Joey and I saw him and stopped, we looked at each other and made sure our knives were in our pockets. Nodding, we took another three steps forward and stopped. Tears swelled in my eyes, I made eye contact with Joey and saw he was doing the same. I couldn't help it, I started crying right there on that sidewalk with my brother.

In that small moment, we had bought into the fear of everyone telling us, 'Those black people are out there to get cha, you better watch out!' Even though, logically, we knew it wasn't true, we had been told for the same thing for so long that we actually started to believe it! We had only been exposed to the South for eight days now, already we were buying into racism? Now, I understood why this kind of hate still existed, people like Joey's little cousin grow up being told the same things we were told for their entire lives. We are all a product of our environment, so how could you not believe that white people are 'superior' in some way to black people if you were told it was true for years?

After this sunk in, I simply had to let the tears flow. I cried not for the pain I was in, nor how impossible our trek seemed, and not even because we had bought into the fucked-up beliefs for the tiniest second; I cried in mourning for all the people that grew up down here in the South that thought this was just the way things were. We only had one option left at this point, we wiped away our tears and without saying a word to each other, walked right up to the man on the side of the street to do our best to make his day. I had to fight away more tears because when the 'man' turned to face us, I was met with the face of a boy. He was 15 years old waiting for his mom to pick him up after soccer practice. To think I was prepared to draw my knife on a kid only a couple minutes ago...

We started talking to him about his day and what he thought about the crazy heat that morning. He told us it made him thirsty so we filled up one of our water bottles and gave it to him along with a couple granola bars in case his mom ran late. I think if that kid would have asked, Joey and I would have given him every scrap of food and water we had in our bag out of the guilt we felt for what we had assumed about him a moment earlier. We left that kid smiling on the side of the road as we walked away trying to hide the tears that had started flowing again.

The best part about experiencing the hate we did for those couple of days is that it brought Joey and I back together. The disconnect between us had healed in the somber mood of that day. It took a couple of hours before we were making jokes again and smiling at the opportunity to achieve something as big as traveling across the country. There was always an undertone of sorrow in what we said for the rest of that day though.

More Sorrow

We were free now; it was time to turn our attention to what was next. Chattanooga, Tennessee was our next stop to visit my cousin, Dawn, who had assured us she was the best chef on planet earth. We had no problems with her threatening to force food on us, by the time we made it the next 120 miles there, we knew we were gonna be hungry.

A man stopped for us a couple hours into walking in a truck with little room in the back. We made a move to jump in the bed, but he absolutely insisted for us to join him in the cab. A little nervous about his excitement to see us, we hopped in to hear yet another sad story. Another veteran, this man was one of the ones that could never quite adapt back into normal life after killing so many people in the east. I believe that he was the second person that picked us up on the brink of suicide.

He was just coming from court where he had lost custody of his son. We asked him why they would do that and he responded, "I just get angry, ya know? I scare my wife sometimes but I don't mean anything by it. She just doesn't understand…" he faded off, mostly mumbling to himself now. Joey and I took that to mean domestic abuse, which we have no tolerance for. This man was different though, he would switch personas from a crazy ex-military wife beater to an innocent and childish one that didn't know how to live in society.

I think he wanted to kill himself because the next thing he did was burst into a fit of excitement about showing us all the weapons he had in his car. He opened his center dash and passed out like 10 different knives to Joey and I, explaining the best way to kill someone with each of them as he went. I didn't want to seem threatening, but he kept insisting that we open and unsheathe them, joking about how defenseless he was now that we had all his knives. Almost as if he was asking us to end his life for him so he didn't have to when he got home. I didn't know how to make him feel any better, so I just listened about how he hated his life.

Wow, the sad stories and experiences just wouldn't end today. That car ride started to get a little sketchy the longer the man told us about how awful his life was, so we had him let us out before we got where we were going. I wished him the best of luck, but that sort of negativity and sadness was not something we needed at that moment. Sadly, that man pulled over and let us out, his childish persona showing itself again. It was hard not to worry about him as he drove off, part of me hopes that he recovered from the pain that was eating him inside out, but the other half of me thinks that death is just the best option for some. It was up to him to decide if he deserved redemption or not.

We walked the rest of the day without rides and made it to a little gazebo that would have been a perfect first date spot. There was a two-person swinging chair, a little picnic table, and hedges that would have made it perfect to sneak away from the super church just down the road. Luckily for Joey and I, it doubled as a hobo shelter just as well. The gazebo gave us shelter for the rain and the hedges hid us from the suspicious eyes of the church. We went to bed that night in peace on the wooden floorboards of our gazebo, the rain singing us a lullaby that promised tomorrow would be better.

Mountains

We had been walking through the flatlands for days. Sure, there were little hills here and there, but nothing on the scale of what we had back in Idaho. Not having mountains around made me feel exposed. There was nowhere to go for solitude, but there were stories we would hear from our drivers and strangers of the Appalachian Mountains. We had learned about them in school, there are the Rocky Mountains on the West Coast, and the Appalachian Mountains in the east. Nothing got me more fired up than the possibility of massive mountains rising out of the ground, so I held out hope that the east still had surprises in store for us.

The morning progressed into afternoon before we saw them in all their glory. Green foliage covered every inch of them like a fur coat, the peaks had clouds swirling around them hinting of the secrets that might lie beyond what our eyes could see; they were magical. While not as big and powerful as the Rocky Mountains, the Appalachians seemed more mystical and seductive. They beg you to look at them in all their majesty, and then the whispering in the back of your mind to come just a little closer start Quiet at first, but eventually evolving into a burning desire to explore what those seductive mountains held just out of reach. Joey and I felt that desire to see our Appalachian seductress in all her beauty, but we had one more thing we had to do first. My cousin, Dawn, would never forgive me if we didn't first stop to say hi in Chattanooga! We only had about 50 miles there, so if we were lucky, we could make it before sundown.

Dawn was more extended family that I knew basically nothing about other than my mom adored her. I understood why when my phone started ringing. Digging through my bag to find the little device, I answered with no idea of what to expect and was greeted with the most wholesome and friendly accent ever. "Jake! Is this you, my man?"

I smiled, "Yes! Is this Dawn?" She squealed in delight on the other end of the line and gave us the best news all day.

She said, "If you can make it within 30 miles of Chattanooga, I will come pick you up myself!" She was on speaker so Joey heard and started nodding my direction vigorously.

"Dawn, that would make our day if you could do that! We will call as soon as we get close." Hanging up, Joey and I felt more motivated now, the difference between 30 and 50 miles was huge, now we just needed one ride to get us there.

Dawn Codename: Our Sunshine

Thinking we would increase our chances of being picked up, we hopped onto a no-name road barely wide enough for two cars to pass each other. That road led us through the most quintessential part of the South we got to see. There were old farm houses around every turn, fields that always seemed to have mist hanging over them, it was not my kind of country, but I can see the appeal to it now. We passed by a fence with bamboo growing over the side and hanging out over part of the road. I walked by first, thinking nothing out of the ordinary was about to happen until I felt a *dink* against my backpack. Slowly turning, I saw Joey hold a stalk of bamboo like a spear preparing to throw another at me. *Oh ho ho, you wanna play that way?* I thought to myself, *It's on...* I bent down quickly and picked up the makeshift spear Joey had just launched at me. Like a standoff in the old cowboy movies, Joey and I stood facing each other, both with a bamboo shaft cocked back in our hand, ready to fire as soon as the other made a move. Our eyes narrowed into focus mode; my hand twitched in anticipation of what was to come...

With a battle cry, we both hurled our spears at each other and the war was on. We turned those bamboo shafts into every weapon imaginable from bo staffs to dual katanas. Reaching down to grab a shorter, more effective piece, Joey broke his bamboo over my backpack. There was no more time to dodge sticks, I went juggernaut mode and started walking through every blow as long as I could guarantee we got the satisfaction of breaking a shaft on Joey. He is good at sword fighting though, I think I only broke one on him compared to the five or so he broke on me. Being that we had walked 10 miles already, our energy tank didn't last long until we both putted out in our war game.

It was a fun couple of miles we had on that little road. Sword fighting, horse petting, we even got a ride from some friendly guys in a massive black truck. Proving that even super hard-core conservatives would lend a helping hand to a friend in need. They even ended up giving me a manual on tick identification that did nothing but scare me about getting Lyme disease.

We had made it to within 30 miles of Chattanooga at long last. I called Dawn and she said she would be out to get us an hour or so. Feeling triumphant, we walked into a gas station to get a candy bar but were promptly kicked out for our looks. A little taken back, we walked through another one of the rainstorms from the hurricane that was hanging out in the Gulf of Mexico. This one lasted much longer than the ones before it, assuring us that the hurricane was gaining ground on us. By the time Dawn arrived, there was over an inch of standing water on the ground. Suddenly, I understood why flooding is a problem in the South.

I would equate Dawn to the likes of a pitbull. Super happy jolly go-lucky fun, but deadly and vicious if you get in the way of something she wants. We crammed into her little mini-cooper in our wet clothes and started on our way to the Tennessee state line. Dawn is a natural conversationalist, it makes sense given her job at a radio station. She is also a very open person that throws all of herself into a conversation, not holding back or apologizing for any of her spunky and exciting personality. She really is one of the pure and good people of the world, it gave me hope to meet her because it proved that people like her are very real.

She pulled over at the 'Entering Tennessee' sign and let me do a backflip to celebrate entering our third state on the trip. The sky opened up and washed away the rain as we hopped back into the car with my fantastic cousin. It was very symbolic in the way that Dawn had just become our sun to shine away the nightmare we had lived through walking out of Georgia.

Luxury

Southern living is a different lifestyle than the one you will find in the Pacific Northwest. Southern living seems to have everything built around comfort and appearance. Fantastic food, shiny cars, the softest of couches and beds, not that I was complaining, Joey and I were amazed when Dawn unveiled our upstairs room to us. She took us up a flight of stairs to a room easily the size of most living rooms. A king-sized bed big enough for at least four people dominated the back of the room. Starting the trek across our new room to the bed, I passed a plasma screen tv and a couple game systems residing in front of a massive couch that looked like it was begging to give you a hug. I put our bags down and heard Joey yell, "Jake! Check this out, man!"

I was confused because Joey's voice sounded like it was echoing. Turning around, I expected to see a portal to another dimension Joey had fallen into (hitchhiker logic). Instead, I saw a long hallway that snaked around a corner further back into the house. *Not as exciting as another dimension, but I'll check it out.* Walking down the hallway, I found my way to what I would describe as a master bathroom. It was not a massive room, it didn't have its own sauna, but what it did have was incredible detail. First off, it was impeccably clean and organized. Second, it was very ornate with flowers carved into the towel racks, the edges of the mirror were decorated with more carvings and insignias that had now led me to believe we were living in absolute luxury.

Dawn politely recommended that we shower before dinner, we took the hint. I went first and made my way downstairs to the most heavenly smell. I had almost forgotten that Dawn had a family wide reputation for being the best chef anyone knew. My mind flashed forward to all the first date meals that I could be cooking for girls with Dawn's knowledge. Running the rest of the way to the kitchen, I found the cheeriest version of Dawn. She was jamming to some Luke Bryan's '*I don't want this night to end*' country music and running around the kitchen with absolute confidence. One of my favorite things to experience is to watch someone do something they are really good at, and Dawn was more than willing to show me that she knew exactly what she was doing. By the time Joey came down, Dawn had prepared a feast of, not only the best food we had since hitchhiking, but some of the best food we had ever eaten. The star of the show was this awesome caramelized banana dessert that was almost orgasmically delicious.

The Land of the Lotus Eaters

'The Odyssey,' an ancient Greek epic by Homer, describes the adventures of Odysseus after the fall of Troy. Sailing for only nine days, Odysseus and his crew were blown off course to a mysterious island covered in lotuses where they met peaceful people that lived in complete apathy and contentedness. The only thing that it seemed they were capable of was a small act of generosity to some of the men in Odysseus's crew and offered the hungry men food in the form of the lotuses that dominated the island. Upon the first taste of lotus, the men were thrust into a state of bliss with little care about anything else in the world except eating more lotuses. In modern culture, we often refer to people in the modern age as lotus eaters when they exist in luxury rather than dealing with practical concerns. Why do I use this analogy? I use it because Joey and I were the lotus eaters, and Dawn's house was the island.

The next morning seemed to come as if no time at all had passed. For the first time since starting our trip, Joey and I slept in until almost 9:30! A big change from waking up at 6–6:30 every other morning. We reasoned our tired bodies must have needed it and dragged ourselves downstairs to find yet another buffet waiting for us, this time the star dish being candied bacon. O god, it was so good. Joey and I are gentlemen though, we could not simply sit and eat while Dawn did all the hard work. As we had done the night before, we cleaned everything Dawn had made for us and put away our dishes. We were surprised when Dawn seemed in awe of having her kitchen cleaned for her. We had thought it was only the simplest thing we could do for her, given the meals she cooked had shown me that I had never understood the world of food before I came to her house.

With full bellies, we got dressed so Dawn could take us on a tour of her work at the radio station and downtown Chattanooga. We were blown away by the personal movie theatre at Dawn's radio station and the beauty of Tennessee. So far, the South had left little that was particularly special in my mind, the bridge we jumped off in South Carolina and the Appalachians in Georgia had both been rather alluring, but compared to Idaho and Montana? We had not been impressed. Coming to Tennessee was a different world, however. Driving on an overpass, we got a look out over the city to see what I would imagine as more of a tropical rainforest climate than a Southern one.

Mist hung over the city, but unlike the fog back home that spoke of dreary days to come, the mist in Tennessee spoke the language of adventure (Joey and I's native tongue). The mighty Tennessee river ran through the middle of the city, wide enough for 2–3 cruise ships to sail side by side. The Appalachian Mountains loomed over the city, the mountains seeming even too sacred for even the greediest people of urban development. It was maybe the only city I have ever seen that nature seemed to be in control of the city instead of the other way around.

We had a hunger to explore now that the subtle hint of racism in the environment seemed to have been left in Georgia. Dawn's mini cooper came to a stop outside of the city aquarium where we hopped out to make the city our own for the four hours that Dawn would be at work. "This city is in trouble now, mwaaahahahahahahahaha," I laughed until I caught Joey looking at me as if he had just discovered his dog was really a cat in a dog costume. I stopped my laugh abruptly and composed myself, "Well… let's go," I coughed out, not able to come up with a better sentence to justify myself.

Joey left a moment of silence hang in the air before laughing and nudging me on the shoulder, "Yeeaaaah…let's do that!"

We spent the next four hours getting kicked out of buildings for looking homeless and breaking into children's history museums before Dawn picked us up and took us to the most incredible ice cream shop I have ever been to. I don't even remember what I got but it was the first time I had ever thought about, "How bad would armed robbery be if I only took ice cream?" I don't really have a filter so that line of thought poured out of my mouth, and yes, it was loud enough for EVERYONE in the restaurant to hear. It got a laugh from Dawn and Joey as well as a couple other tables, but as you can imagine, the person taking orders at the counter gave me some crazy eyes that said, 'I'll fuck you up dawg.'

I responded in eye talk, 'Bitch, I'm a hitchhiker. You got nothing I ain't seen yet.' Not sure how you put a Southern drawl on an eye stare down, but I'm damn sure I killed it.

My stare down only lasted a couple seconds before we finished our ice cream and left back to Dawn's place in the mini cooper that I grew to love so much. On the way back, our inner lotus eater really started to come out in force. It was so easy to be content all of a sudden, that hunger that drove us to fly across the country was being buried under the layers of luxury that Dawn had so kindly exposed us to.

We had made a list of rules when we saw how great Dawn was treating us:

1. Help out as much as possible, make beds, fold clothes, do dishes, anything that helps.
2. No TV or video games, part of the trip was supposed to be separating from the time wasters.
3. Spend as much time with Dawn as possible.

Upon coming home, we broke rule number two like we had forgotten we made it. Dawn's newly married son came home to have another 5-star meal with us of enchiladas. After dinner, we dragged our swollen bellies to the couch and plopped down in front of the TV with Dawn. We justified breaking rule two with rule three, that was almost okay and going back, I would probably make the same decision. However, the fact that we had broken a rule already was evidence of us drifting into a life of comfort without ambitions.

Leaving Lotus Island

After some self-reflective talk earlier in the day, Joey and I had realized that we were becoming content and had to break the cycle. Before we went to bed that night, we made Dawn promise to take us out of the city early in the morning. She had told us she could drive us out if we were ready to go by 6:30. *Easy,* we thought, we had been getting up at that early the whole trip without an alarm. Just to make sure we weren't going to fall back into our lotus eating ways, we blew up an air mattress and slept in the living room on the floor.

Rolling out of bed the next morning was peaceful, we had slept so well. I reached over and looked at the time that read 6:36 a.m. *No! Had we missed her?* We ran to the kitchen bouncing off walls as we went. There sat Dawn at her dining table, phone in hand and happy as a clam.

Dawn: O good! You're up, I warmed up our leftovers from last night for you.

Me: Uh... Weren't you supposed to leave at 6:30?

Dawn: Don't worry, hun, I called in and said I would be in a little later. I had to make sure you were well rested! Feel free to stay another night if you need to.

She said this all with the jolliest bounce in her step, not faltering at all now that we had made her late. Now she wants us to stay with her again? She was so perfect that it was tempting to stay another night all of a sudden. We snapped out of it, Joey and I knew we had to start on the road again or we might never leave.

Me: Dawn, thank you so much for everything, but we really do have to get on the road to make it home in time.

She protested, she felt she needed to make sure we were well fed and taken care of, but we had to insist. "Thank you, Dawn, but we really must leave." Succumbing to our request, Dawn drove us 30 miles to make sure that we were well out of the city and had a chance to keep going. With a final hug, we said our goodbyes to one of the most generous people we have ever met.

"Thank you!" we shouted as she drove away.

How Will We Ever Go Back?

With a deep sigh, we sat down on the sidewalk next to a Dollar General to wait for the 8 a.m. opening time so we could fill up the food stocks. It was there that I remember one of the most meaningful conversations that I had with Joey. Sitting on the hard concrete with a slight drizzle raining down on us, I felt at peace. It was the most rewarding thing to know that this was all us now that we had escaped lotus island. No one was taking care of us even though we were about to walk miles and miles, may or may not be picked up, and even questioning if today could be our last day alive. Whatever came, be it great or tragic, it was all us.

That moment was enlightening for me, we felt so right in the new lifestyle we had adopted that we sincerely questioned how we were ever going to go back to living as normal members of society. We are all born with an innate need to fit in that goes back thousands of years to when people roamed in tribes. If you were kicked out of that tribe, you were dead. Now, in the modern era, we still have the instincts to fit it even though it is not key to our survival. Joey and I had each other and the road, there was no more pressure to get a job, no more bringing in groceries for my mom, no more struggling to climb the social ladder.

We honestly questioned if we wanted to go back anymore, we could get jobs here and there to fund our never-ending adventure, we wouldn't have to worry about the fakeness that so many people are filled with, everything out here was 100% real. If someone picked us up, they were real to us. It didn't matter if they wanted to kill themselves or just do a deed of kindness for their fellow man, THEY WERE REAL. Being hungry was REAL. Walking until our feet bled was REAL. Back in our old lives, however? I had spent my entire life pursuing basketball, nothing more than a game. Joey had fallen in love twice and been kicked to the curb for being too nice both times. Making friends in our old lives took weeks because you would have to wait until you needed them and wonder if they would show up. The TV I watched with my dad, the video games I played with my brother, the computer screens I looked at in college, NONE OF IT…could come close to the real-life journey that we had found hitchhiking for the past 485 miles. How could we ever go back?

Punishment

We set off at a slow and thoughtful pace to start. Taking in the absolute beauty of Tennessee as we climbed a mountain road that followed the Tennessee River northwest. We felt so good that we took a detour to an underground waterfall (yeah, we thought it sounded badass too). On the way up the mile-long road, it started hailing. Yes, it was summer time in Tennessee and it was hailing thanks to the hurricane that was still shooting off mini storms northward.

A tree provided us some cover as we smiled at the beauty of nature. How lucky we were to be able to experience such a natural phenomenon. Upon arriving at the waterfall, we found that it was $70 for a tour guide to show us that cave. We didn't have that money to spend, all I had brought enough for was a food budget. That is something I really hated. It was humiliating for me to have to choose between fun and food. It was good to live in frugality for us on this trip, but I never wanted to be turned away from something I really wanted to do again because I really needed the money for something else. I swallowed my pride and retraced my steps with Joey to the winding mountain road that followed the river, we were finally lost in the Appalachians.

The forests were thick, and on such jagged slopes that a person would be hard pressed to walk without using their hands to drag themselves up the steep slope. No cars were stopping to pick us up today, maybe partially because hitchhiking is punishable by law in Tennessee. We walked for 20 miles on that road through some beauty that rivaled even Idaho's natural landscapes. Even when there is such beauty around you though, 20 miles up and over a mountain is no easy feat. We were dragging our feet by the time we made it to the next town where our whole trip would change.

Santa Clause

This was maybe the smallest Southern town we had been to yet. Only 10 minutes of walking and we had made it through the entire town to find the convenience store. I waited outside with our bags due to the suspicious looks we kept getting from the locals. Lucky for me, I had stocked up on pop-tarts this time. Calories and sugar were all I wanted after those 20 miles. I was beat and ready to fall asleep right next to that cute little picnic table until I heard Joey smiling as he walked back (yes, we were at the point where we could hear each other smiling).

Joey: Jake, there's this big guy over there who looks like he's kind of the head of the town. He just told me that there is a train yard, 'Where trains go slow enough for a man to hop on,' just over the bridge over there.

Me: No way… Should we do it?

Joey: Fuck yes, dude!

Me: Hahaha okay. I gotta meet this guy though.

I sauntered over to the little store looking for, "A big Santa Claus looking dude," according to Joey. I didn't find him right away so I loaded up on another five chocolate pop-tarts and filled my water. On my way back out of the store, I saw him, *Good God, it is Santa himself.* Sitting on a bench on the porch sat a very large man with a dirty white beard. He was easily 300 lbs. and smiled his cheeky red smile at me when I came through the door. Sticking out his hand, he introduced himself as Billy Ray or something along those lines (basically, the most trustworthy name anyone could be given).

He told me, "Awww, son! I dun met y'all's friend but second ago. He tells ya there's a train yard?" His eyes locked onto mine, unable to look away, I nodded my head slowly. Billy Ray leaned forward in his chair and said, "Son, dat train is gonna be in Nashville tomorrow morning. Y'all best be on it."

I walked back to Joey with the same happy grin on my face. We had to trust this guy, right? He was bestowed with the most typical Southern name I could ever think of and looked like Santa Clause! It would be a betrayal of our childhood instincts if we didn't listen to what the man had to say, right? We talked it over for a minute before deciding that we were going to be on that train tonight.

The Second Longest Night of Our Lives

Joey and I made our way across a bridge to the railroad tracks. It wasn't long before a train rounded the corner going slow. "This is it!" I yelled to Joey. We hid behind the concrete pillars that supported the bridge above us until the engine passed us. Hopping out from our hiding spot, we ran to the moving train. Upon getting there, however, we realized that the train was picking up speed, it was too risky. The good news was that we knew it had stopped at some point to be going that slow, that meant we could walk down the tracks and find where the stopping spot was and hop on.

The black railroad rocks made walking hard but we went anyways. Several more trains passed us as we walked another couple miles down the tracks, all going just a little too fast for us to justify hopping on. So, we walked…and walked. We ran up and down those railroads tracks deep into the night. At one point, we decided that if a train was going to stop, it would wake us up, right? It was pitch black outside, all we had were our headlamps to guide us to a steep edge along the side of the railroad. There, on the black railroad rocks, Joey I would sleep only 10 feet away from the tracks waiting for a train. Laying down next to each other, we closed our eyes, not that we slept. I dozed off a time or two, but would always be woken by another train that was moving just a little too fast for us to jump on.

Hours passed and the rocks were not getting any more comfortable. There was always that one that had to stab into your back in the most annoying spot. The only comfort we had was that this was about as bad as it could get and we were okay, then it started to rain. Not a downpour this time, just enough to get us wet. "This isn't the right spot, man," Joey said to me after a couple minutes of rain on our face.

"*Sigh* You're right. I think the spot we had earlier was better." Standing up, we began to backtrack back the way we had come at 2 a.m. in the rain.

There is a certain level of amusement that comes with situations like that night on the railroad tracks. I think that maybe it is pride in ourselves, knowing that things have backfired as badly as they could and yet still, we keep believing that it is going to work out somehow. We had now been up for somewhere between 19 and 20 hours, it was raining, our legs were barely functioning through the punishment, we were cold, and had already depleted most of our water supply. Still, we laughed in the dark. *What could be better?* I thought to myself.

We're on a Train, Baby!

It seemed I had just fallen asleep when Joey woke me again with excitement in his voice. We had worked our way back into the woods and set up our tent when the rain became too much the night before. Now, here it was at 5:30 a.m. and Joey was yelling, "Jake! Jake! Get up, man, we gotta go!" I had forgotten where I was completely.

"Where are we going?" I asked him.

He started shaking me, "Dude, there's a train! A TRAIN!" Suddenly, I was awake, our only goal from the night before had rushed back to me. Peering out the mesh tent window, there it was in all its beauty. A train stood still on the tracks right in front of our tent! We couldn't let this opportunity pass.

We took down the tent in record time in our hurry and sprinted out of the woods as fast as our half-dead legs would carry us. Finding a car that wasn't dirty from all the coal, we threw our bags on and found a place to sit right above the connector between the train cars. It created a little bit of a cage around us with the ladders that went off to the top of the car, this would be our luxury suite, but the train wasn't moving. We waited just long enough to get suspicious before we heard the unmistakable crunch of boots on rock. There was someone else walking on the railroad tracks! We didn't know if it was going to be another train hobo or what, but we did not want them to see us.

To better describe the scene, we were at a spot where two railroad tracks ran next to each other. The trains had stopped because there was another train that needed to pass on before the tracks merged back together to create one. On this particular morning, the two trains had stopped next to each other, creating a crazy amount of hiding spots for your inexperienced train hobo.

The steps were coming from in between the two trains and getting closer. Joey had his arm dangling in the middle of the tracks in plain view of whoever it was that was walking towards us. I stood to run but Joey motioned me to sit back down. "He saw me," Joey mouthed. All my hopes of riding a train deflated instantly, at very best they would kick us off the train, but we had heard stories of railroad crews beating the shit out of train hoppers like ourselves. *We might have to fight our way out,* I thought to myself. Joey had withdrawn his arm, so we waited for the footsteps to reveal whoever was making them.

The steps belonged to a construction worker in a safety vest. He walked up so close to that I could hear his panting in a struggle to carry his large body over the uneven rocks. I held my breath, it was time to face our demons, but the man kept walking. He didn't even look up! He was so close that Joey could have used his head for an armrest! Joey and I looked at each other and held down our laughter. Before we had time to say much of anything, we heard more footsteps ahead. Stashing our bags where no one would see them, we embraced our inner Assassin's Creed. We separated like we had been born to hide on a train. I climbed up the ladder to the top of the railroad cars to serve as lookout while Joey watched out for anyone that would spot me below. We avoided their wandering eyes for a couple of minutes until they all walked back to the front of the train. Joey and I came back to our spot and waited in anticipation. Then we felt the first lurch of the train, WE WERE GOING TO NASHVILLE!

We Aren't Going to Nashville?

It really felt like we had traveled back in time to the age of discovery. One must realize that when you travel by the highway system in the United States, there are routes to take you to every urban development you want to go to. The train tracks are not quite the same. Yes, the railroad still takes you through big cities and the occasional small town, but their purpose is different. The train destinations are cargo yards that are usually on the outskirts of the city, and to get there they take the fastest routes. This means avoiding as much human development as possible for long stretches.

Within the first 10 minutes on that train, Joey and I had entered a different world. One minute we were following a highway that never seemed to leave the masses of people, now we were alone without a human structure in sight other than the train we rode upon. We followed the river upstream along the edge of the steep Appalachian Mountains on our left elated, Joey and I couldn't stop smiling. Here we were on day 12 and we had now reached hobo level 3!

We were completely separated from the outside world except for one thing. We still had our phones. While we didn't have internet wherever the train was taking us, we did have our location. Zooming in until we could see the railroad tracks, we discovered we had just taken the ultimate gamble. As Billy Ray had promised, there was a track that went to Nashville. What he didn't tell us or probably didn't know, was that the tracks split, one leading to Nashville, and another that led down to the heartland of Alabama. Joey and I are not religious men, but if there was a time on our trip that a prayer might have gone a long way, it would have been right then. We arrived at our crossroads still going 40 mph, leaving us no option to hop off and try again if the train took the Southern route.

We could see the split in the tracks ahead of us. I did my best to ground myself, knowing that whatever way the train decided to go, I would make the best of it. I still wasn't prepared when our train turned South. I briefly considered jumping off even though we were still going 40 mph, but realized there was nothing I could do to control what had become our new Fate. We had no choice but to roll with the punches now.

Alabama

Alabama was a State we had hoped to avoid after the Georgia racism incident. I don't know why, but all the stories we were told in high school of civil right movements and the horrible crimes that were committed against the people who participated seemed to all come from Alabama. We didn't even want to imagine what we would experience now that we were in the state we associate most with racism. The only way we didn't lose our minds was that we kept reminding ourselves of the facts:

1. Most people are good.
2. Everyone wants a good person to talk to.
3. Alabama being a racist state was a huge generalization.
4. We will survive.

The train finally came to a stop about 160 miles away from Chattanooga and another 200 away from Nashville in an unreasonably famous town named Muscle Shoals, Alabama (for some reason every, time I think of that town now, I associate it with Larry the Lobster from Spongebob the TV show). The first thing we had come to realize about Alabama was that it was much hotter than the cool Appalachian environment that we found ourselves in only six hours ago. Hopping off the train with the sun beating down on us, we found ourselves on a service road alongside the trainyard that led to an asphalt road we could see in the distance. "At least there will be cars there," I said to Joey.

The heat was so great that distance could not be accurately predicted because of the visible heat waves radiating through the air and altering our vision. What we had thought to be no more than a mile soon turned into a three-mile walk that we had very little patience for. We were angry with the whole world now, nothing was going to cure that. A friendly railroad worker did his best to cheer us up with some mini water bottles and asked if we had been riding the train. Not wanting to blow our cover, we lied that we had just walked across a field and finally found ourselves on this service road. The man seemed nice, but we couldn't risk getting in any legal trouble with such limited funds.

It was another two miles to the nearest grocery store to fill up our water bottles, but we were exhausted. We hadn't been able to sleep on the train and ran out of water a couple hours ago. In this heat, the lack of water was brutal. We trudged our way to the store, our bodies moving much slower than our minds compelled them to. After filling up and restocking our food supply, we were lucky enough that a man in a pick-up stopped in the Dollar General parking lot and let us jump in. The guy was unnaturally happy to have us in his truck, but it did help me come back to a somewhat stable spot in my mind. *Just a little set back,* I was telling myself. We can still make it to Nashville.

Our new driver was so excited to show us around his town that we ended up driving around in circles for 15 minutes before he even asked us why we were hitchhiking. Joey and I didn't care too much, it was so nice to be in an air-conditioned car for a couple of minutes. I answered him, "Sir, we're hitchhiking back to Idaho."

Before I could elaborate at all, the man cut me off and blatantly said, "Aww, you're fucked." Okay; wasn't exactly the sort of motivation we were looking for. He went on to tell us that nobody gets out of Muscle Shoals, people hate hitchhikers here, you're lucky I stopped for you.

Alright, that's enough, I thought. Joey had gone quiet, so it became my job to get us out of the car. Despite the fact that the guy had only driven us a total of one or two miles towards Nashville because of all the things he wanted to show us, I told him that, "We just need a place to sleep tonight, do you know of any parks?" He dropped us off down by the river, insisting on walking us down to a spot he thought would be good to sleep. I had no more toleration left for this guy. I told him that we would sleep at the first spot I saw and put my bag down. Joey did the same and we stared at him until he left. At least that guy was gone, now we just needed a spot to sleep for the night.

The Bridge Talk

I felt sick from the day we just had. Life had taken every hope we had and crushed it. If we could just get some sleep, maybe we would feel better. We couldn't sleep in the middle of the sidewalk where we had put down our bags. Unfortunately, there had to be a better spot somewhere in this little park. Joey was still quiet, only breaking his silence with the occasional grunt that prodded me to keep looking. We walked by a couple little hobo tents in the woods, but Joey and I were not about to sleep with some trash bag looking for inexperienced adventurers to prey on. We found our way down to a huge walking bridge a couple minutes later on the underside of the highway. It was the nicest spot we had been since we started walking, as good a place you could ask for to sit down and feel sorry for ourselves.

Joey told me something once that will always stick with me. He said, "You know why we're friends, Jake? It's because the optimists and realists need each other. The optimists are the kite in the clouds and the realists are the man on the ground holding the string. Without the kite, the man would never look up at the sky and see what was possible. Without the man, the kite would stray too far into the clouds to never be seen again." Joey often saw himself as the realist, or the man on the ground while I was the kite in the clouds, the source of his inspiration. I think he couldn't have been more right looking at the relationship we had before this adventure. Now, the optimist vs realist dynamic we had thrived on was breaking down.

We were mad, more than that we were helpless to do anything about it. We took off our sweaty shirts to let them air out and sat across from each other on the walking path under the bridge. Before I could say anything, Joey spoke and said something along the lines of, "This sucks. I just want to be back with people that actually care about me." What could I do but agree? He was right, this did suck. People would drive by us in trucks with empty beds and one person in the cab and not pick us up. We could deal with that, what we could not deal with was the mental toll of watching trucks that would drive by us, with their beds filled with trash. Joey said to me, "They are literally treating their trash better than they are treating us." Now, that is a very logical conclusion to come to, and he was right. The problem was that I wasn't supposed to agree with Joey on things like that; I was the optimist! I had to be excited to do anything and give everybody the benefit of the doubt even when I knew it was completely illogical.

Sitting there on that bridge listening to Joey talk, I couldn't refute his logic with blind happiness anymore. I had to listen and nod because he was RIGHT. We weren't prepared for this level of apathy where we were going. The vast majority of people didn't seem to give a shit about the person next to them. And those who did pay attention? They were the ones that usually swerved at us. Searching for those few that would take time out of their day to help someone in need was an excruciating process. The more we talked, the more we realized that it wasn't only the people around us who were mean and apathetic, we were becoming the same way.

The longer we were in this environment, the harder it got to be nice to a stranger. We were beginning to see everyone as our enemy instead of potential friends and that scared me. The Jake and Joey that sat on that bridge were not the same two kids that had started this adventure, not even close. I was ready to put my head down and tough through it, but the more I listened to Joey, the more I realized the change that was happening to me. I pulled out my phone and took a video of myself, then scrolled all the way back to when we had first started this trip to compare them.

Watching the first video, you can see me just after slicing my foot on a piece of glass on the first day we started walking. The old me is a cocky clean-shaven kid, he looks so carefree and happy to be walking along the side of the road despite his cut foot. Fast forward to day 12 and you see a much tanner and realer Jake. Both good things, but under that you can see a hardness that wasn't there before. That hardness was coming from my changed view of people. They were no longer all unique individuals, now they were just tools that I could use to get where I'm going.

There were no more real connections with the people I was talking to. I was just playing to their egos so they would tell me all about themselves so Joey and I wouldn't have to answer the same questions again.

'Where are you from?'

'Why are you here?'

'Do you have a family?'

'Are you in school?'

'Aren't you scared?'

Every single freaking conversation was a play on those same questions. They are good questions, but we had often already answered them four times a day to the gas station lady, the person that would stop to hear our story and drive away after we told it, our last ride, again and again. Out of respect, we had to answer them, we couldn't just say no after someone had been kind enough to pick us up. I didn't mind having a great talk with somebody, but what sucked was that we knew they didn't really care! They were just asking the same questions they had been taught to ask and except for a very select few like Caitlyn from Wyoming, Bobby from South Carolina, Pam and Bob from Yellowstone. Other than a short list of people like them, no one could relate or care about our journey.

It is once again one of the times I am happy Joey was there for me. Without him, I would have fallen asleep in the park and woke up the next day to try again. The consequences of that decision could have changed my character in ways I can't even imagine. Instead, Joey and I bought a very short plane ride from a local airport to Nashville, TN.

What Did We Do?

Although it was getting late, we discovered that we only had 5 ½ miles to the airport. The plane was leaving early so we would have to be there early. The only way to insure that was to start walking again. I have never not wanted to keep walking more in my life up to that point. Still, we stood back up like the badasses we are and started walking.

If there was one thing the guy earlier had said that was true, it was that Alabama does not take kindly to hitchhikers AT ALL. We had about three people swerve at us that were only half serious, never enough to get a real reaction out of Joey and I. Then one car slowed down behind us to scream when they got as close as they could without us knowing they were there. That one was enough to get us both to jump as they peeled out and sped past us laughing. You would think that would have been as mean as people got, right? No one else was going to get any meaner than those dirtbags. You are wrong. As darkness fell, a man decided he would swerve at us and mean it, the mirror on the right side of his brown and tan truck hit the brim of my hat on his way by, I was lucky I didn't have my thumb out for a ride or he would have taken my arm off.

That behavior was something I would never understand on the trip, I still don't to be honest. When you look back at those cars that would swerve, it really was a threat to kill us. They were just letting us know that at any time, they could turn their wheel just a little sharper and end our trip right there. Why would someone intentionally target a helpless person on the side of the road? We were young kids of 19 and 20, we did our best to look somewhat presentable every morning, brushed our teeth, fixed our hair etc. What did we do to make someone threaten to run us over with their car? Joey and I had accepted how at mercy we were to the situation around us and didn't even break stride when the brown truck flew by. Any reaction would only give them what they wanted. It did prompt me to walk in the ditch next to the road after the truck was well out of sight, but Joey kept on walking where we had almost been hit a moment ago. Almost as if that in the case someone did hit him, it would be revenge on all those who had wronged us after our 12 days on the road, knowing that whoever did it would go to jail.

The brown truck was the last one to swerve at us that night. We made it within a mile of the airport and found a patch of grass outside a little league baseball field that looked comfortable. Not wanting to be breaking and entering, we threw down our sleeping bags and ate a dinner of Chef Boyardee ravioli and some Cheez-It I had picked up earlier. It was hard to complain, other than the couple mean people from earlier, this was a pretty good night. We were looking out over a picturesque field with a single tree in it and eating our fill. All in all, it was better than most nights until I felt that first mosquito bite.

The first must have told the rest where we were because we were swarmed by the thickest mosquitos I have ever seen. This day would just not end. We lathered on the Deet I had brought as fast as we could and curled up inside our sleeping bags. If you stayed still too long, however, they would bite you through the bag! It was another one of those comically awful moments on the trip that lead to very little sleeping. We were almost thankful when it started to rain a few hours later and drove the mosquitos away, but now we had to figure out where to sleep so as to not get soaked.

Coincidentally enough, we were on Joey's home turf, a baseball field. He had played baseball for years, so I trusted him when he led me over the fence to one of the dugouts with a tin roof. Yes, we were sleeping on concrete again and it wasn't completely dry, but at least we would be able to sleep here. I woke up to a tingly feeling on my neck. My first thought was that Joey was trying to wake me up with a piece of grass or something so I rolled over away from it. The feeling didn't go away though, in fact, it multiplied. There were little tickles all over my body! I stood up and discovered I had gone to bed directly on top of an ant nest last night. "Whhhhyyyyyyyy!" I took off my shirt and pants to shake them all out and emptied my sleeping bag. They were everywhere! Joey was very entertained at my yelling and dancing to get the ants off. It took a while, but finally I said, "Okay… I think that's all of them." With a breath of relief, I reached over to put my shoes on. Well, it seems the ants had not only decided to crawl inside my sleeping bag as their home away from home, they had also turned my tennis shoes into their vacation condo! "Ahhhhhh!" I yelled and threw my shoe at Joey who was laughing his ass off.

There Are Always Good People

Alabama had not been good to us so far. We found a wall plug in to charge our phones and sang some Johnny Cash to get us back in a good, almost melancholic mood. As if nature knew that we hadn't had enough bad luck for the night, she started hailing on us. The hail was so hard that we had to take shelter again to avoid the bruises we would get from standing in the open. It was exactly what we needed, Joey and I played games of running out in the hail and rolling in the wet grass before running back under our roof. Faith in humanity was still at an all-time low, but our personal mood was now officially restored, giving us enough energy to hike the final mile to the airport.

The hail had stopped but we could see the sky turning dark grey again. We picked up our pace and made it to the airport 90 minutes before our plane was set to leave, only to discover that our plane might not even come today due to the severe weather they were experiencing in Atlanta where the plane was coming from. "Heh," I sighed. There was no use in complaining, we were just going to be here for a long time, better than walking in the hail at least. As luck would have it, in the most miserable time is where we finally met some good people.

Something about a shared bad experience brings people together. Everyone in that tiny little airport had just had their day ruined. The worst luck among them being an older couple who were going to be missing a friend's wedding because of the storm. That was sad, but they still smiled and told us stories about the people they used to know that were hitchhikers. I'm pretty sure the husband hinted that he was one of the Freedom Riders from the Civil Rights Movement in the 60's. They left after a short while and we met a man that had just ordered a pizza. He was too distracted to really talk to us, being on the phone almost the whole time we sat next to him. He unknowingly convinced Joey to call and order us a pizza as well because we were starving once again. The man's pizza arrived shortly after Joey ordered and we had to leave because the smell was too good.

When we came back, we caught a glimpse of the man leaving in a black truck, but his pizza was still there. Sitting down next to it was torturing ourselves, we didn't know if he would be back to claim his lost untouched pizza so we 'guarded' it for him. Our will lasted for a couple of minutes before I was shoving slices down like there was no tomorrow. Then Joey's pizza arrived and we had a full-on feast in the airport of two whole pizzas. That was the first stroke of luck. Soon after, we met some of the most wonderful people.

We first saw them when one of the most adorable little girls I had ever seen in my life came up and introduced herself to us. My heart melted, I was ready to give her what was left of my pizza before her mom rounded the corner and ran to save her daughter from the two hitchhikers that were trying to feed her daughter stolen/borrowed pizza. While we did get off to a spotty start, it didn't take long before the mom introduced herself as Nekiar. She was shocked to hear our story and some of the things we had done the past 13 days on the road. Her daughter wasn't quite as interested and running around the airport like she was to be a future ninja warrior. That made me like Nekiar even more, she wasn't restraining her daughter from having fun like so many parents I had seen on this trip. Joey, Nekiar, and I talked for over an hour about everything under the moon. Despite being a mother, Nekiar was only a couple of years older than us. Our age allowed us to connect much better as she told us about her family and past relationships, even the job she worked so hard at to support her daughter. Upon leaving, she labeled us her spirit animals and told us to be safe. It is always just the little things in life that mean the most. Nekiar's support and tiny bit of caring gave me back the hope I was going to need to finish the trip.

Nashville, Baby!

We touched down in Nashville after embracing our sky hobo. We were three hours late, but it didn't matter, we were HERE! The only city we had wanted to see for the entire trip was now in front of us. We were two small-town boys that had never liked the city, what made Nashville so exciting, you ask? We were huge country music fans that had dreams to sing in every karaoke bar we could get our hands on. Needless to say, we were so stoked to be here.

Nashville lives up to its name, folks. The place is amazing, every bar is owned by the most famous of country music's superstars. The Florida Georgia Line bar, Tim Mcgraw Saloon, 'Ol Red's Bar owned by Blake Shelton. Going inside, you would find beautiful girls dressed in their cowgirl boots and short skirts. On stage, there was always a band singing county's most famous songs like *'The Thunder Rolls'* by Garth Brooks or *'Barbeque Stain'* by Tim Mcgraw. The only downside was that I was hoping to hear some Shania Twain or Carrie Underwood, but never saw a female band the entire night.

The only female singer I remember was a wife and her husband that killed an old bluesy song in a karaoke bar later in the night. Joey and I were set to go on next with 'Red Solo Cup' by Toby Keith but we couldn't match that! Joey's confidence was unshakable though, he had been told how awesome his hat was the entire night and now he was ready to go. Even though I didn't know all the words, Joey pulled me right up on stage with him to lead us in the best version of that song that has ever been sung. Joey even sat down on the edge of the stage to make sure a lonely woman knew he was singing straight to her. I did my best to keep up with the guy, but he was on fire and totally in his element. I sang the backup vocals for most of the song while Joey took the spotlight and hell, he deserved it. We ended the song to receive some of the biggest applause I have ever gotten in my life! We bowed together and stepped off stage like newfound country stars that had just landed a record deal.

I don't think we have ever been better friends than in those couple of nights in Nashville. When we weren't singing karaoke or listening to the greatest live bands I had ever heard, we were exploring what daytime Nashville had to offer. The Country Music Hall of Fame was number one on the list. Everything from Elvis Presley's custom car to the thousands of hit records on the wall was amazing. The thing that stuck out to me most, however, was the circle chamber where all of the Country Music Hall of Fame inductees were held. It is rightfully that last thing you see in the tour as it leaves you breathless. Each inductee gets their face carved into a brass plaque and hung around this massive room. For how big it was, there were not that many people who had actually made it. I loved seeing names like Reba McEntire, Dolly Parton, Alan Jackson, Garth Brooks, George Strait, as well as some names I didn't know would be there before I toured the building, like Elvis Presley, Patsy Montana, Willie Nelson, and Loretta Lynn. It was incredible to see all those names that had made history. Walking around that room made me ask a question that would later inspire me to write this book with Joey, "If they all changed the world and made history, why can't I?"

Never Ride Greyhound

We enjoyed Nashville for the two nights we were there and, other than a crazy drugged up guy with a gun that wanted us to come outside with him, it was some of the most fun we ever had together. Enough was enough though, it was time to head over to the dreaded greyhound bus that would take us to Denver. Shouldering our backpacks, we walked eight miles through Nashville to another completely new experience at the bus station.

It makes sense that Greyhound is a lower income kind of transport, we expected some interesting people, but nothing we couldn't handle. I knew that place was sketchy when I had to remind the kid loading the bags under the bus not to forget one when he went to close the bus up. A little on edge already but still wanting to make up for the kid I had almost pulled my knife on in Atlanta, I passed up an empty pair of seats for Joey and I to sit next to one another and chose two older, rough looking gentlemen in the back of the bus to accompany BAD CHOICE.

The one guy ended up being able to hold a conversation, but insisted on giving me life advice that made no logical sense. "Yeah, mun, ya know, just get that shit dun. Feel me, brother? I used to be just like you and now...well, ya, I used to be like you and den I got into the drugs, ya know. Yeah, it was tough." Even if the man was the richest dude in the world (which he was not), I couldn't have understood what it was he wanted me to do anyway.

I grew tired of that conversation pretty quick and turned to the other older man on the other side of the bus. He was a Rastafarian-looking guy, but he did not look good, and he smelled, but I decided to give him a chance anyway. I leaned across the aisle and asked him, "Yo, man, where you headed?" He looked up from shuffling through his bag like he had just been caught in the middle of a bank robbery. It took him about five seconds before he realized it was me talking to him and smiled the most incoherent smile I have ever seen to this date. With one finger, he motioned me to come closer as he bent back down into his bag. I was not about to get any closer to that guy, but I didn't want to be rude so I watched him fumble through his bag until he seemed to find what he was looking for. "Hah!" he exclaimed with a voice I could only imagine would belong to an elder tribal witch doctor and pulled out a vial filled with some sort of liquid in it. He examined its contents until he was satisfied, then offered it to me.

Crazy witch doctor man: Cu buuii?

Me: I'm sorry what? What is that?

Crazy witch doctor man: Cuu buuuuuiiiii?

Me: Woah, chill, bra. I don't think I want that.

He seemed to think very hard now and stared at the seat in front of him.

Crazy witch doctor man: Rwwwwrrr…eh rwwwwrrr…

Me: Uhhh…you okay, man?

He looked at me like I had just said something profound and coughed at me before diving back into his bag. The breath from his cough smelled like death and almost knocked me over, but I had to be polite so I watched to see what he would offer me next. Coming back up from his bag, he had held a second vile that was filled with a plant that I recognized; marijuana. With a prideful expression on his face and thrust both vials at me this time with words I understood.

Crazy witch doctor man: Tifftyyyyy!

Me: What? Fifty?

Crazy witch doctor man: Da! Fifty dolas.

Ahhh, I see what's happening now! I'm part of a drug deal, I thought to myself… *Wait, I'M PART OF A DRUG DEAL?*

Me: Uh uh… Uh no, nooooo. Yeah, no, thank you.

The crazy man was persistent and kept thrusting the vial of marijuana and what I can only assume was heroin at me saying, "Tifftyyyyyy!" I had to say no like six times before the witch doctor stood and went to the bathroom. Seconds later, I could see and smell marijuana smoke pouring out of the seams of the bathroom door.

It didn't seem Joey was having a much better time up front, he was doing his best to protect a very large gentlemen from the fat shaming he was getting from three people sitting around him. Joey had switched into the seat next to him and was doing his best to handle that and another lady in a wheelchair that seemed to be living in her own world, switching back and forth between crying and laughing. You would think that was all that happened on that bus ride, but no. All this was within the first 40 or so minutes on the bus.

Those couple things were not the most shocking that would happen to us on that bus though. For the 26 hours we spent sleepless on that bus, the most surprising situation we dealt with was the self-proclaimed gangster. I finally worked my way back up to Joey's seat, saving me from being part of a drug deal in the back. I was relieved until I overheard the stories the guy in front of us was telling like he was the baddest mofo that had ever walked the Earth. He was talking about his sexual escapades of how he had sex with three generations of women in the same house. First having a girlfriend that took him home to meet her mom, promptly had sex with the 'way sexier' mom, and moved onto the grandma just a few days later. It couldn't stop there though, we were on a Greyhound after all! The self-proclaimed gangster then started telling us about how he had fucked with an older woman in her baby's room while the baby slept and was now riding this Greyhound to Colorado to do the same thing. *Awesome,* I thought to myself. *Fuck you, Greyhound.*

Now that I am older and wiser, I have had time to reflect on my Greyhound experience and would like to more properly represent it. Greyhound provides cheap travel to those in need… Nah, just kidding, my opinion hasn't changed at all, I still fucking hate Greyhound.

The Rocky Mountains

Hitchhiking is an eye-opening experience. As I have highlighted already, life had never been more real for Joey and I when we were walking along the side of the road for hours in the heat with no way to pass time. It was filled with very long periods of discomfort, pain, and hopelessness followed by a tiny good thing happening that we would perceive as one of the most incredible things that ever happened. It really made us appreciate the little things in life. The 26-hour drive through Kansas was a prime example.

Kansas would have been an amazing place to visit about 500 years ago when the bison still roamed the vast grasslands that used to make up the Midwest. Now, Kansas has exterminated just about everything that used to live there with the exception of a few nature reserves I was not lucky enough to see. Looking out from our Greyhound, you would never guess that this land was once full of life. Looking across those vast plains, I could see no life, with the exception of the occasional cattle farm, and no diverse grasslands that used to span this state. At the end of the day, I'm happy those farms provided me with food on my table, but it does make me question if the prices paid to feed the human race are worth it when you see such an altered landscape.

Kansas lasted for ages and I was not impressed. The good part about it was that I knew there had to be something good waiting for us in Colorado. I thought they were just a shadow in the distance at first, but the closer our bus got, the more I became sure that we were not looking at any shadow in the distance, those were the Rocky Mountains! Although we were still 1,100 miles from Sandpoint and hitchhiking would actually take me about 1400-1500 more miles before I made it back, I felt like I was already home.

Adventure
Loving Nature and Life

Back in the West

Almost had a heart attack when we couldn't find our bags buried under the other luggage. I thought back to that kid that almost forgot a bag in Nashville, if my bag got left on the other side of Kansas, I was going to go full rage monster on Greyhound. There was no need to worry though, our bags turned up and we headed off into the mile-high city of Denver, Colorado nestled beneath the mountains. Now that we were here, there were so many possibilities on what we could do! Colorado has adventure in every direction, how were we possibly going to decide? The only thing we knew is that had to work our way towards Gallatin National Forest where we were going to look for treasure! That meant we had to take a rough heading west, but other than that we were free to explore.

Looking off towards the mountains that were so close was almost a tease now, neither Joey or I had the patience to walk 12 miles to see them so we tracked down an Uber to take us. I always thought that being any sort of taxi driver would be a scary job. When our driver showed up and ended up being a younger woman, maybe around 30? I asked her about if she had ever had any scary experiences. She quickly replied that she was a black belt in jiu jitsu and those of you who don't know, that basically makes her a real-life ninja that could kill you if she looked at you too hard. Fully convinced that not even Chuck Norris would be able to beat her in a fight now, I settled in for what I can only imagine were the safest 12 miles of the entire trip.

She dropped us off in a town named Boulder. Perhaps the most famous town in all of Colorado, Boulder is home to the most beautiful college campus I have ever seen. It makes sense with the almost $60,000 price tag that comes with it if you want to attend there. I was utterly fascinated by this place; there was a kombucha store on every street corner and at every house there were college kids hanging out on their porch that were more than willing to socialize. That was a town I could have spent a couple of days in, I even brought the idea up to Joey, but his mind was in a different place.

Joey was a man on a mission, he reminded me of 'Focused Jake' from back in the South. He was not even willing to stop at a coffee shop when I saw a beautiful girl that I wanted to go say hi too! Now, that was a level of strange I had never seen from Joey. I didn't worry about it though, whatever anger or emotions he was feeling had nothing to do with me if he felt that strongly. I let him blaze a trail up the street to the mountains, but I was in no hurry. As far as I was concerned, I was already home now that the Rocky Mountains towered over me. I took my time talking with strangers and doing a couple round off backflips in the grass at the university. It was fun, but ultimately, I had to keep up with Joey. We were brothers, after all, I couldn't choose Boulder over him.

The Ascent

When I managed to catch up with Joey, he had stopped in a beautiful little park where he threw down his bag and made it very obvious he did not want to socialize. That was fine with me, I had been there enough times to know that I had to give him space, plus I had a park to explore! There was a nice little stream that flowed down from the mountains above that was freezing. Dunking my head in that water made me feel more carefree than I had on the entire trip. I re emerged from the cold water with a whole new mindset, I wasn't worried about getting home or seeing my family, or even really finding the treasure. I just cared that I got to explore my extended home of the west.

After some yoga with some college girls and gathering some local knowledge that there was a lake in the mountains we would see if we went west to Nederland, I made my way back to a much happier Joey. He was smoking his e-cigarette, something I had now come to terms with, and seemed like he had cooled off. He told me he was just tired from the bus ride and even offered to walk back into Boulder with me. I almost took him up on it, but all I really wanted was my partner back. We had hitchhiking to do so I didn't care about Boulder anymore, now it was time for the ascent into the mountains.

Now came the question of, 'How does hitchhiking the West compare to the South?' We decided that the narrow winding road of the mountain probably wasn't the best road to compare the two, due to it being hard to stop and not get hit by traffic. Regardless of that fact, two hours into walking, a Colorado man picked us up. He was another former hitchhiker and had just recently moved with his wife to Nederland, 'The paradise in the Mountains' as I like to call it. Nederland sits at over 8,000 feet above sea level, after coming from below 1,000 feet in all the Southern towns we had been, Joey and I were feeling the thin air for sure. It's good we were already in phenomenal shape, otherwise it would have taken at least a week to adjust to the altitude.

Something we had noticed about all the little towns we had been to so far was that you almost always see the same people. We walked inside the one grocery store and saw a beautiful girl at the checkout, walking out we saw the group of kids making trouble that Joey describes as 'Hick-thugs.' The same kind of grouchy old man hobbled past us in the parking lot. There was only a single noticeable distinction, perhaps it was us and our confidence in being back in the West, but people here were so much more open! It wasn't like you couldn't make friends in the South; most people are very nice there. What was different was that, with the people in the South you had to break down the prison walls that guarded every emotion and thought they had before they would open up to you. Here in the West, however? It felt more like white picket fences that guarded people, I could vault right over that sucker and make friends in no time.

Freshly stocked on food, we climbed higher into the mountains outside of Nederland. The burning in our lungs was almost addicting the higher we climbed. We were looking for something, we didn't know quite what it was, just that we needed to keep going until we found whatever we were looking for. Up the winding road of the mountain we went, no matter how much our legs screamed or how hard it got to breathe. I was beginning to wonder if we were just masochists, but then what we were looking for revealed itself to us at long last. It was the peak of the mountain above Nederland and our camping spot tonight.

Hopping off the windy road, we snuck through a yard and up the mountain. Fueled by obsession to see whatever view awaited us, we stopped at just under 9,000 feet elevation and looked out to see our past on the right, and what would become our future on the left in full view. We saw the vast plains that our bus had driven across that morning and the edge of Denver where it had dropped us off. The town of Boulder stared up at us from thousands of feet below as if it was lifting us up, and last, the reservoir/mountain lake that sat on the edge of Nederland where we had just come from. Looking to the left, the vast forests spanned over the endless mountains. Somewhere past all of that was our goal, where Forrest Fenn's treasure awaited us. Exhausted but inspired, we hung our hammocks instead of the tent for the first time and let the wind rock us to sleep.

Hippie/Mayor/Businessman

We woke up with the sun lighting up the horizon. There was adventure out there to be found! We were both in a fantastic mood now that we had recovered from the 26 hours of sleeplessness on the Greyhound. We packed up all of our stuff and decided that we needed to get our blood pumping before we started our long day of walking. We hadn't quite made it to the peak the night before and that just couldn't be tolerated. Throwing off our bags, we ran up the mountain faster than I thought possible. Our legs had grown used to having an extra 40 lbs. weighing them down. Now that we were free from the weight of our bags, we could run like the wind.

The altitude slowed us down a little, but all in all I felt like I was in the best shape I had ever been in. Our powerful legs pushed us up the rest of the mountain, the birds' morning songs cheering us on as our breath became ragged. Reaching the top, we both knelt over on our knees, breathing hard. The view was even more beautiful than we had hoped, but there was too much energy to still be spent to fully enjoy it. We went back to our roots on the way back down the mountain, jumping off little cliffs doing 360s off rocks, we were to do what we please, when we please, how w... *Crack* "Aaahhhh!!" Joey yelled behind me. *That's what we get for having fun,* I thought. I turned around to discover that Joey had just come an inch away from falling through the top of a fort built into the side of the mountain.

"Woah," we said in unison. It looked like we had just stumbled upon a doomsday preppers hideout. The fort looked years in the making; from the side you would see an abrupt right angle dug straight down into the mountain that would have taken an enormous amount of work to get through the rocky soil. The excavated dirt had been used to make elevated walls on every side of except for a small entrance that you could crawl in and out of. Old logs had been drug over the top of the whole thing that gave it the feel of a cave underneath. It would hardly have been waterproof, but the owner had fixed that problem with a plastic tarp over the logs and weighed it down with large rocks and as a final touch, they had thrown dirt over the top of the tarp to make it blend in with the landscape.

We did a real quick look around to make sure there wasn't some old Vietnam vet hiding behind a tree with his bowie knife, but this thing was too awesome to do any thorough surveys before we crawled through the opening and pretended we had just been dropped into Red Dawn. "Joey, the Russians have taken over EVERYTHING. We are the rebellion...will you fight with me, my brother?" He returned my deep eye contact and grabbed my arm in the brother's embrace.

Looking deep into my soul, he told me, "Jake...I'm with you until the end, brother." We fell into character like we had been practicing for an Oscar the entire trip.

Me: We've been here too long, only a matter of time before they find our position... Wait, I got it. We should set up base camp in Idaho!

Joey: Roger that, brother. We have only one form of transportation since the Russians took out our vehicles though.

Me: What's that, soldier?

Joey: We are going to have to travel via the civilian population as hitchhikers.

Me: Understood, I'll alert the rest of the rebellion and follow your lead on this one. O and, if we die out here to the Russians, I'm proud to die with you brother.

Joey: It was my honor to fight alongside you against the Russians. They may take our lives, but they will never take our freedom!

Me: I think you switched movies there, bro.

Joey: Yeah, yeah, whatever, I've seen way more movies than you anyway.

Nothing like little sarcasm and laughter to start our first full day in the West. About four miles in, we met our first ride for the day. A bus pulled off the road that had 'Peace Rides' sprayed on the side of it. Running up alongside the tiny bus that looked like a blue version of the magic school bus, the sliding door opened to reveal a husky older man with a 5-star white beard. I don't remember his name, but I will always remember that German Shepard's name that I became best friends with named Gandalf. Yes, like Lord of the Rings Gandalf, aka one of the greatest role models ever/sexiest grandpa alive...that Gandalf. With a 100 lbs. furry Gandalf at my side, I had the best ride I could ask for.

Although I was far more enamored with the dog than the man that was driving us, our driver did have some stories to tell. Joey and I soon came to find out that we were riding with the former mayor of Nederland. That was surprising news given the man's appearance. He wore a tie dyed purple and white shirt, his hair was long and unkept, and he was driving in a bus that I had assumed he probably lived in. Well, it seems appearance is not always the best indicator of someone, as our driver had a number of claims to his name.

He had been elected mayor amidst one of Nederland's worst financial times that he could remember. They had a huge amount of debt to pay off that it seemed no one was capable of taking care of, so he ran on the platform of getting Nederland out of the red and back into the green. That was enough for the people of Nederland and he was put in office. He called himself an, "Untraditional politician," because he actually followed through on his word and did exactly that. Before his term was over, Nederland was over a quarter million dollars in the green.

We were impressed, but he was still voted out of office the year after his financial feat. He made it very clear that he was done with politics and his business approach to them. He took those political skills to start his own company instead and named it Peace Rides, hence the name on the side of the bus. Now, he used his bus to shuttle drunk tourists to and from town for a price. *What a genius,* I thought. We talked business strategies and cuddled with Gandalf until he dropped us off at almost 10,000 ft elevation.

Colorado is so Awesome

We said a heartfelt goodbye to Gandalf and hopped off the bus. Stepping out into the thin mountain air was exciting, we could see where the road reached the top of the mountain and began back down the other side. We had every intention of sprinting every bit of that last mile or so up the mountain, but as it always does, reality slapped us in the face. It couldn't have been more than three bounds before we were wheezing from the elevation. It seemed that not even Joey and I, with our tree trunk legs and iron lungs, couldn't adapt to elevation changes that fast.

Walking was all we could manage as we finished the next couple of miles. The cool temperatures allowed us to walk for much longer periods of time than we had been able to in the South. Lucky for us, we didn't have to walk that far before an awesome old van with only one sliding door pulled off the road in front of us. Wow, two rides before 12? That never happened...

A man rolled down the window with a smile on his face, "Hop in, guys! I can take you to the next town!" We were thrilled since the next town was a lot further than we wanted to walk. We slid open the side door to find a thrilled little girl that couldn't have been older than five. That represents the massive change in the environment we had just gone through. Nobody in the South would have let us sit alone in the back of the car with their little girl, now we were being asked to? I recoiled at the thought of us being trusted with a little girl, it almost seemed like it shouldn't happen.

When we were in the South, we only got rides if the person in their car was alone. Only once did someone pick us up that had kids in the car, but we were in the back of the truck with a clear and defined barrier between us and them. It was like that for so long that I almost started to believe that I was a strange man that wasn't to be trusted. *I could be dangerous, right?* It was like my inner voice was talking to my less friendly personality I developed in the South.

This was the predicament I now found myself in staring at this little girl in the back of the van who desperately wanted to play with us strangers on the side of the road. Joey hopped right in, to her delight, and without a moment's hesitation, they were deep in conversation about unicorns or something of the sort while I elected instead for the front seat with the dad. The van was set up almost like a taxi, we were only connected to Joey and the little girl in the back by a little window that only gave us the occasional squeal of laughter while Joey and the little girl had their play date. The dad laughed, "You guys have no idea how much of a favor you just did me. That little one has waayyyyy too much energy to be stuck in the back by herself today." I was blown away by this man's trust in us, we hadn't been trusted for over two weeks prior to making it to Colorado. I didn't say anything for a moment as I thought back to what life was like for me as a five-year-old.

My mom and dad used to send me to walk the dogs around the block as a little kid, I despised it, until I found out that the doggy treats that the neighbors would give the dogs weren't half bad. After that, I couldn't get enough of them! I would stop by every neighbor's house in the neighborhood to get three dog treats, one for each of the dogs and one for me. I also remember hitchhiking for the first time. I couldn't have been over eight when an old lady took me back to my driveway when I got tired of walking home. Hiking in the woods alone was also a common occurrence as a little kid. The most relevant experience, however, was when my dad brought home a train hopper to eat dinner with us as a kid and I loved it. I realized I was now that train hopper for that little girl and her father. I was back with people that think like me, I didn't have to be scared, I didn't have to watch my back, I only had to look back at that smiling father with a smile of my own. "Happy to help," I told him.

We wound our way down through steep Colorado mountains until we stumbled upon the weirdest town I have ever been to. We had reached our driver's destination, so we gave the little one in the back a high five and stepped out of that awesome old van. Now we could clearly see what this weird little town had to offer. It sat in the middle of massive gouge in the mountain, sheer walls of rock towered over us on each side, telling the story of past mining in the area. Underneath those walls stood almost equally imposing casinos 10+ stories tall that seemed fit for even the largest of cities, but there was no city to be found. The closest town was Nederland with a population of around 2,000 people, not near enough to justify building these massive casinos. Why were they here?

Wandering around yielded no answers to our questioning minds, this would have to be a mystery left unsolved. Not another mile down the road, a young couple picked us up in an awesome little Subaru. I looked over at Joey, "I love Colorado, man." We hopped in to find the interior of our new ride was decked out for rock climbing. A light chalk dust coated the interior of the vehicle and carabiners were thrown about the car. Well, it just so happens that

Joey is an insane rock climber as well, and these were his people. Not 30 seconds into our car ride and Joey was talking to them about v-10s (some sort of rock-climbing term) and laughing with our new friends. After we rightfully earned their trust, we got our answer on why there are random casinos as far away as you can get from the big cities. It seems the idea of casino vacation in nature appeals to enough people that multiple massive casinos can survive over 50 miles outside the city of Denver.

Freeway Hitchhiking

Our rock-climbing friends took us through a long tunnel that opened up into a beautiful canyon drainage that had an at least two-hundred-foot flat face that served as perfect rock-climbing turf. We all got out of the car and said our goodbyes, as our drivers started up the steep hill to rock climb. Joey was looking longingly at the wall, he so badly wanted to go climbing with them, but the desire to find the treasure outweighed his want for rock climbing.

We set a mean pace today, fueled with motivation from the three rides we had already, we were not going to stop anytime soon. It was good that we were ready to walk, because rides were not coming anytime soon. We were in mining territory now, which meant the only traffic was 16-wheelers taking rocks filled with precious minerals to and from the mine. This meant no rides on that windy day; no semi was about to pick up hitchhikers on the side of the road. On and on we walked until we saw a terrifying site ahead of us, the freeway.

Joey and I had come to have a special relationship with the freeway by this point in the trip. Hitchhiking on the back roads was what we had spent the majority of the trip doing so far. There was usually not much distance to be gained when we were picked up in a 40-mph speed zone, but we could count on at least a ride or two every day using that system. We had only tried hitchhiking on the freeway once, only to give up after waiting for hours at an exit, hoping someone was going to pick us up. We much prefer to have at least some of our future in our own hands instead of handing it off to a stranger with no more than a few seconds to decide if they wanted us in their car. Joey equated getting a ride on the freeway to winning the lottery, the pay-off was massive, but the chance was very low. We had no choice but to buy our ticket and play the waiting game now.

In front of us sat the last portion of the active mine, beyond that, our road joined the massive freeway that took up all the space between the two steep mountains in front of us. We couldn't walk on the side of the freeway, not only was it extremely dangerous, it was also illegal. "Well, shit," I heard Joey mumble next to me.

I sighed, "Yeah...yeah, man," I told him. We made our way to the spot where our road curved under the huge bridge of the freeway and joined on the left side of the road. We settled in for a loooonnng wait under that bridge, this was a mining road, used almost primarily for giant semis carrying minerals, and as we had quickly learned, semis do not pick up hitchhikers.

We waited about 90 minutes with no success until we saw a car slow down! "Yes! Someone is finally going to pick us up!" I stood up and waited expectantly. The car slowed to a snail's pace as it approached us...before it sped up and took off up the on ramp.

"Ah, just wanted to get our hopes up," I said to Joey. He didn't respond. I think we had lost the ability to take anything personally at this point. I turned around expecting to see my brother in arms still sitting in the shade on the side of the road. Instead, I turned to see him scrambling up the steep rock wall that led up towards the bridge above us. That is something I love about Joey, when life rejects him, he doesn't just take the punishment, he takes action.

I walked over to the steep wall; Joey had already yelled back down to me over the noise of the traffic that there was a turnout we could wait at. That was great news, waiting on the side of the freeway instead of the exit would give us a way better chance of being picked up because of the difference in traffic. I was no rock climbing extraordinaire like Joey, however. He had just climbed up that wall with his 50 lbs. backpack on, no problem. I could have reached up and let Joey take my bag, but my pride was on the line here. Refusing Joey's services, I inched my way up the wall. Although it did take me much longer than Joey, I made it up without any falls or slips to find ourselves in a similar position to before. Playing the waiting game again... Lucky for us, people don't seem to hate hitchhikers in Colorado. In the South, I wouldn't have been surprised if it was a six-hour wait. Fortunately, we were in Colorado, it was only 40 minutes before a red car stopped for us. "God, I love Colorado, Joey."

This is How Hitchhiking was Supposed to Be!

One of my favorite parts about getting picked up is the unveiling of the driver. Everyone was different, old, young, man, woman, but they all tended to be pretty similar on the inside. If they were a cake, the recipe would look something like: three cups openness, one cup of a love for life, two tablespoons generosity, sprinkle in a sense of wonder, one teaspoon curiosity, don't let it settle, instead bake it right away in a pan made of something...different. That would pretty well summarize the people that picked us up. Our new ride was no different, her name was Joan.

A sweet older woman, Joan was cautious when stopping. I don't blame her; I can only imagine we looked rough after our last couple of miles walking through the mining zone. I don't know what it was about us, but people just trusted us after they stopped. It didn't take but 15 seconds before we were in Joan's car and telling our story. Lucky for us, she was headed almost 80 more miles to the next town of Vail. We chatted nonstop for the whole time and left enough of an impression on her that she didn't want to drop us off. We only convinced her to let us out on the condition that I took her phone number just in case we couldn't find another ride.

As she drove away, we discovered perhaps the most brilliantly named business in all of the US. The City Market grocery store. When Joey and I saw that, the first thing our minds went to was a year-round, indoor farmers market. "What an honor to serve the local businesses, and what great food!" Well, it just so happens that The City Market is Colorado's version of a chain grocery and that they are in every single little town. It took a while to put those pieces together, but when we saw the third store, our suspicions were confirmed. That whole time we thought we were supporting the locals...oops.

Joan had dropped us off at what would become one of my favorite towns in the world. Vail, Colorado. Colorado in general is magnificent, but Vail had everything you could ask for. Driving on the highway, it would be easy to miss. The highway runs east and west on a hill just north of Vail that sits in the valley bottom. Looking down on it, Vail almost seems to be squished between the hill the highway runs on and the incredible mountains that tower over the town. The buildings are modern, but built with a sort of adobe looking exterior that allows them to blend into the landscape.

I have never been more taken with a town. It did look very expensive, so perhaps it was good we didn't get the opportunity to stay very long. A jeep squealed to a stop beside us and the man inside yelled, "I'm going to Avon! You coming?" I was tempted to say no just to hold onto my precious Vail a minute longer, but Joey made the decision for us.

"Yes! Thanks for stopping, man," he said. You could tell the driver was in a hurry, as soon as we closed the door, our heads were plastered against the headrests on our seats.

He was a handsome young man that you could have mistaken for late 20s even though he was almost 40. He was one of those people that you just can't help but be enamored with because of how happy and excited he was about living. I have since deemed his personality type one of the 'untouchables.' It doesn't matter what you do or say, he is going to have the best day of his life every day. Underneath his happy exterior, there was a sense of calm and sturdiness as well. I wanted whatever that was, it took me a while to figure out what made him that way, but looking back, I understand. His calmness and quiet confidence came from the true belief that he could conquer or handle anything that he wanted to. Compare that to me, I was super happy and excited around people, but I didn't have that same sureness in myself. It was that sureness that was taking our driver on a date with a 25-year-old girl in the next town and why we were in such a hurry to get there.

We got another taste of his confidence when we arrived in Avon. Unlike most drivers, he didn't ask where we wanted to get out, he just stopped on the side of the road and said, "Okay, good luck." I could almost feel the masculine energy radiating off of him, there was no doubt that he was the alpha male in that situation. Joey and I couldn't do anything but get out and watch as he drove around the corner to meet his girl. I can only assume that she was among the most beautiful of women to go on a date with that man.

My New Favorite Town

As I watched that awesome man drive away, I slowly came back to the present moment. I know I had just decided that Vail was my favorite town in the world, but Avon was even better! It had all the cool buildings and architecture that Vail had, all the cool coffee shops and local cuisine, it was amazing. I was having trouble figuring out which one I liked more until I found the gold mine; Avon has its own lake! Well, more of a reservoir, which isn't near as cool, but we could swim in it! Walking around the edge of the water, we even found a sand beach that seemed to have the whole town sunbathing on it. Joey and I couldn't pass this opportunity to jump in the water! Much to the beach goers dismay, we stripped down to our underwear and dove into the not quite warm enough to swim in water.

That was a place that we could have stayed at for a while. There was a perfect little tree covered area where we could set up our tent, we could have some local breakfast in the morning, maybe even go for a hike! I brought up the idea to Joey, but he was a man on a mission. He had already discovered there was a bus route that ran to the next town and had decided that we needed to be on it. I argued with him for a bit, but in the end, I didn't want to spoil the amazing mood that Colorado had put me in. I never thought I would be so reluctant to put my pants back on in a public setting. Somewhat annoyed, I followed Joey back into the heart of town to find the bus station.

It seemed Joey and I had switched roles. Whereas I had been the A to B boring stick in the mud for the few days in South Carolina and Georgia, Joey had now taken that role and was more focused on getting to the treasure area as soon as possible than having a good time. Had I been thinking further ahead, I might have been able to see that the different reasons for which we had come on this trip were starting to show. This wouldn't be such a good story if I had seen our differences coming though.

Local Knowledge

We had already come over 100 miles today, but we were far from done. We found the bus easily. Expecting it to be late, I ran off to see if I could find a local gift shop so I could remember this amazing place, but for the first time in history, the bus actually showed up on time. I sprinted back across the cool sidewalks with fish and grizzly bear tracks carved into them, making it back just in time before the bus opened its doors. Next stop, the town of Dotsero.

While I hate Greyhound with a burning passion, I have developed a certain fondness for the city bus systems around the country. I find it funny that everyone puts their headphones in at the same time when they sit down on the bus, almost like it is a requirement to be able to ride. I don't know if it's fear that makes us hide from one another, or perhaps just not being satisfied with the people we find ourselves in the presence of. Whatever is, I was not going to have any of it. Joey and I struck up conversations with the people closest to us and greeted the newcomers as they got on.

I got so sucked up into my conversation that I became completely unaware of where we were going on the bus. Not that I really cared, I could have missed our stop and gone back to Avon without much of a complaint. Joey was on his game though, just as the bus was pulling away, Joey realized that we were about to head east again. "Jake, this is our stop!" I barely heard him; I was engrossed in a conversation with a woman three times my age about what cooking ware she uses when cooking fish. "Jake!" Joey said even louder and smacked me on the back of my head, "This is our stop!" I snapped into action. Luckily, we had included the bus driver in our earlier conversations and made him our friend. He stopped at a non-scheduled drop off point and made sure to give us all the best tourist spots on our way through the famed Glenwood Canyon that was apparently only about four more miles up the road. With our new local knowledge, we thanked him and skipped off the bus happy as a couple of clams. Seven more miles, and we might find ourselves at a spot of local legends, 'The Hanging Lake of Glenwood Canyon.'

A Man on a Mission

Damn, if Joey put his mind towards it, the guy could be a professional walker no problem. His long legs and determination to make it the last seven miles to Hanging Lake tonight made for a killer combo. He was walking so fast that I was having to alternate between jogging and walking to keep up with him. He probably would have waited had I said anything, but my desire to rise to the challenge of keeping up kept my thoughts to myself.

I had fallen behind again, I kept expecting him to need a break soon, but he just kept going. Gathering myself, I turned my walk into a jog again and muscled my way up to him. I finally told Joey I needed some water, forcing him to stop for a minute while I took off my bag and enjoyed my first glimpse of the huge canyon in front of us. That awe was replaced with shock and anger as I retrieved my water from my bag. The solar charger cord was gone. No doubt having fallen off from the running I had been doing to keep up with Joey. The anger passed almost as fast as it came; it was my own fault. I briefly considered going back to look for it, but looking back at the long road. We had just sped walked 2 miles which took all hope out of that possibility. We would have to hobo even harder if we wanted to keep our phones charged and our video journal alive.

Glenwood Canyon

Glenwood Canyon is about 12 miles long as the crow flies. Carved out by the Colorado river who knows how long ago. You can see the flooding events imprinted on the Glenwood Canyon's walls 100 feet up. To even imagine a river flooding that big was beyond me, much less experience it. Interstate 70 is the only road that runs through the canyon, which makes hitchhiking almost impossible. That ended up being a stroke of good fortune as it forced us to experience the entirety of the canyon on the bike path that runs underneath the highway. That way, we got to explore more nooks and crannies than the highway above allows.

Joey finally slowed down as the walls of the canyon rose up on either side of us. I think he finally accepted that we weren't going to make it to Hanging Lake tonight. Instead, we found a rest stop a couple of miles into the canyon that was a perfect sleeping spot. There was still a little light left, so I took my opportunity to explore some of the grandeur of the canyon. I had always heard stories about the power of the Colorado River. It was called 'The Mighty River' by early explorers. Something with the power to carve out the Glenwood and the Grand Canyon should only be named as such. I worked my way down to the river bed expecting to see the mighty river stretched out in front of me, but that is not what I saw.

It is hard to imagine the impact people have on the environment until you see some of the areas of the world first hand. We hear about that Pacific's 'Trash Island,' we hear about rising sea levels, we know they are a problem, but never really relate it to our own world. In this case, I had heard about the drought in Southern California, Nevada, and Arizona. I knew that they were taking some 10 million-acre feet out of the Colorado every year to supply their water demand, what I didn't quite understand was how that affected the river that breathes life into the Southwestern United States.

I could hear the river flowing. I walked through the thick brush on a small trail. I knew I was getting close to seeing one of the mightiest rivers in the US! My heart beat faster, my adrenaline pumped harder, I even felt my mouth go dry in anticipation for what was sure to be a life changing moment! And there it was. Confusion was the first thing that came to mind, I had to question if this was the right Glenwood Canyon for a moment as I stared, awestruck at what simply couldn't be the Colorado River. Crestfallen, I looked over a river that did not deserve the title 'The Mighty River' for a second. It was smaller than the Clark Fork River I grew up on back home. Standing on that bridge, I could have thrown a rock across the Colorado to hit the opposite bank. Looking up the canyon walls, you could see the river levels where it used to be. The high floods could be seen easily, but if you looked closer, you could see the history of the river. It was a memoir of the past when the Colorado used to run free.

Standing on that small bridge I was frozen, unable to tear my eyes away from the horror laid out before me. For the second time, that hitchhiking trip brought me to tears. I could only imagine the giant fish that used to lurk in this river. For example, the Channel Catfish used to grow to sizes of 70+lbs., now it is rare to catch one over 30lbs. I couldn't help but let my mind wander back in time to when the river monsters used to rule this canyon, but those times were long gone

In Case One of us Doesn't Come Back

Wiping the tears out of my eyes, I slowly meandered my way back to Joey. I was going to tell him of the awful scene I had just laid my eyes upon until I discovered what Joey had done. I couldn't stay sad as I saw that Joey was cramming himself in between the wall of the bathrooms and the vending machine to get at the plug-in so he could charge his phone. Even funnier was that it was one of the Coca-Cola machines with all the lights on the front, I laughed when I heard a crack of electricity and saw all the lights go out on the machine. I faintly heard Joey exclaim, "Fuck yeah!" in success and laughed even harder. This had to be vandalism or something, right? It was way too funny to even consider stopping him as Joey wormed his way back out from behind the machine with a grin on his face that would have made you believe he just slayed a dragon single handedly. Knowing that our phones were going to be charged by morning, we settled into sleep on the cool grass.

We had every intention of falling asleep, but something was nagging at me. One of the philosophies I live my life by is that whenever things are going bad, there is always good right around the corner. That law swings the other way as well though, whenever everything is perfect, you can always count on something bad right around the corner. That day had been so good and pure that I knew it had to swing back the other way soon, and when hitchhiking, there is no way to know how far the other way that pendulum was going to swing. It hit me that night in Colorado that tonight could be our last night. It wasn't fear, just a sadness about what effect it might have on the people I knew. Knowing that things were bound to get much, much harder very soon, perhaps even hard enough that one of us might not come back at all. Joey and I made a set broad promises to each other in the case of only one of us coming back.

First, we promised that if one of us were to die the other would speak at the funeral of the other to tell about how much fun this trip was.
Second, no matter how much protest we got, if one of us were to die, the other would finish the trip in honor of the other.

Then we got into the more individual promises. I gave Joey a list of tasks that I would want him to do in the case of my death. Of course, visit my family, outside that I made him promise to leave my body in the forest so I could be recycled back into nature. At some point after finishing the trip, he needed to go to Missoula with a carrot in hand for Morgan, 'The girl worth fighting for,' I had chosen to help me get through the trip. I had hardly ever talked to her except when I had given her a carrot that she was thrilled about. So funny how the male psyche clings onto the tiniest things like a smile or a flip of the hair for so long.

Joey had only one task for me, he had said everything he wanted to say except to the girl that had ghosted him. Because of that, he asked me to simply find and talk to her. Joey had met her at a blood donation bus where she took his blood. Joey told me that he knew that he liked that girl instantly because his head and mouth stopped working in unison; the words just couldn't seem to come out. I had never seen that happen to Joey. Just listening to the story, I could feel the spark that he had felt when talking to that girl. Unfortunately, she had a boyfriend. Even so, Joey still spent as much time as he could with her, holding hands, talking for hours on end. One of those beautiful relationships that don't come very often. Sadly, it wasn't to last. She blocked Joey on everything when they got too close because she didn't want to cheat on her boyfriend. It hurt him, but he gave me her name in the hopes, that if he were to die, two of the people he cared about most would be able to share stories about him.

We are the same way, Joey and I, we are both so alone…together. Neither of us had anybody but each other on that trip. Even upon returning, there is nobody that can come close to understanding the way my thinking has changed since that trip except for Joey. It's great to have that sense of connection to someone, but at the end of the day, you know that your best friend is on a different path than you. We will always be there for each other, but we cannot always be with one another. The only way that dream will fulfill itself is in the case that we both find a woman to create that connection with. Whoever those girls are, they have a tall order to fill because neither of us settle well. Until those perfect girls come around, it is a lonely existence for the both of us.

Joey and I woke at the same time freezing. We had come from the South, where it was easily 50 °F on the coldest of nights. Now it was dropping into the low 40s here in Glenwood Canyon and we did not have the clothes for it. The first hints of light peeking over the canyon walls suggested it must be about 4:30 a.m. *If I can get another hour of sleep, I will be good for the day,* I told myself. Rolling back over to nuzzle up against my bag, I closed my eyes to be interrupted by Joey's voice.

Joey: Hey, Jake.

Jake: Errrrruuummm?

Joey: I think there are sprinklers here.

Jake: Hehhhh. Even if there were sprinklers, Joey, don't you think they would wait until the heat of the day to turn them on? We're good, man.

Not 15 minutes later, a sprinkler head popped up directly underneath of me. *God dammit,* I cursed in my head, *I hate being wrong.* I tried to avoid eye contact with Joey for the next couple of minutes as he joined me out of sprinkler range. He had set up his hammock the night before, which meant while I was running out of the sprinklers, Joey was having to untie his hammock amidst the 'ch-ch-ch-ch-ch-ch-ch' of the sprinklers soaking him.

Eventually, he joined me on the bench I had found. With a plop, he dropped his soaking hammock next to me and steamily sat down. I could feel his eyes burning holes through the sleeping bag I was using as a hood. I could tell he was trying not to laugh too, but he sat in silence and waited. I couldn't take it anymore, I finally turned to him and said, "Okay, you were right."

"Yeah, no shit, Sherlock," he said with a chuckle. Seeming satisfied he knew that us being wet was my fault, he stood up and went to go huddle in the heated bathrooms to dry out his wet clothes.

I was doing my best to not to start laughing so I shouted after him, "Sorry, Joey!" Without even turning around, he threw up his middle finger and didn't break stride while leaving a water trail dragging his wet hammock behind him.

Hanging Lake

The morning drama now over, it was time to get a move on so we could see the famous Hanging Lake. Apparently, it was one of the hardest hikes in the area, but only about a mile long. How could that be? Upon reaching the beautiful trailhead in full sunlight, we discovered why that was. "O, it's straight up," said Joey out loud. The tourists laughed as they walked by us up the trail. We ate some granola bars and prepared for the accent on a nearby picnic table until we saw a lady that looked about 70 pass us and start up the trail. O, uh uh, we were not going to have that for a second. I started up the trail determined to make it all the way to the top without stopping.

This was where I excelled out of the two of us. Joey's long legs made him hard to keep up with on the flat ground, but the mountains? This was my territory. Without stopping, I rocked that mountain using my screaming legs as motivation to make it to the top. I won't lie, it did feel good to know that I could outpace Joey going uphill because he was such a freaking fast walker on the flat ground, this was far more of a personal challenge. I felt spoiled with all the rides and busses we had been taking. It was time to prove to myself that I could still push through a little pain. In no time, I was at the top of the mountain waiting for Joey before I went around the last corner to see the lake in all its glory.

I shrugged off my 40 lbs. bag and held in my laughter as an older gentleman exclaimed to his friends that I wasn't breathing as hard as they were. I hadn't thought about how strong Joey and I had become in a long while. I had just done the Hanging Lake trail in one go despite the 40 lbs. weighing me down. I thought deeper about it for the next couple minutes until Joey arrived at the top with his 50 lbs. bag still on his back. He was breathing hard, but like me, it only took him a minute or so for his breath to return to normal. I think that was one of the proudest moments of the trip for me. The knowing that Joey and I could outpace about 95% of the human population with weight on our backs felt amazing. There is an awesome confidence that comes with knowing how strong your body is.

With my chest puffed out, Joey and I walked the last hundred feet to the most beautiful lake I had ever seen. Some sort of special microorganisms that live in that lake give it a special sort of cyan blue glow even in broad daylight. The unusually high mineral content in the water also allows for a multitude of strange plant species to grow, none of which would I ever guess were native to Colorado. While the tourists enjoyed the lake, Joey and I set to work on something far more important, cleaning up the trash problem.

While the lake itself is gorgeous, the outskirts of it had all sorts of plastic bottles and wrappers. This tragedy was something we could change though. I didn't have much room in my bag to put trash in, but Joey reminded me how important it was to make room when preserving something beautiful by climbing down a small waterfall to retrieve a bottle well off the beaten path. It was inspiring and I doubled my efforts to clean up as much trash as I could. It feels good knowing that we left that lake in better condition than when we arrived.

We found some friendly strangers to leave our bags with and continued further up the trail to see more of the mountain. Just above the lake is an amazing waterfall, but not just any average waterfall. A rock overhang hung above the walking trail easily 200 feet tall. Towards the bottom of the huge rock face was a single hole where the water from a tunnel spews out with enough force to launch it horizontally to the ground for a moment before it is brought down by the force of gravity in a massive arch. We were awestruck, along with the forty or so other people standing around us.

We let the awe pass and went to work with our analytical minds, Joey with his physics brain and me with my ecological one. I still don't understand how he did it, but from Joey's calculations, the water was exiting the tunnel at approximately 30 mph. I did my best to come up with some sort of smart and calculated thing to say as well, but words escaped me. Whatever had made that water travel in such a bizarre way is truly a marvel of nature. To understand it would only be taking away from the magic that we got to witness in that bizarre waterfall.

Back to the Grind

Maybe three hours or so passed at Hanging Lake while Joey and I told our story to passersby who asked. We even made friends with a guy named John that we still keep in touch with to this day. Though we had originally planned to stay at least half the day at the lake, we were men in motion. There was no way we could sit still for that long while our legs were itching to get going again. I don't know how we did it, but we ran down almost the entire mountain with our massive bags on our backs playing follow the leader. We were copying each other doing 360s off huge rocks and jumping over the creeks that had bridges over them, because who needs a bridge, right? In a flash, we were back at the bottom of the mountain, back to the grind.

Over the next 10 or so miles, we walked through some of the most gorgeous territory we had seen thus far on the trip. The notable occurrences outside of the awesome landscape were when we made it to another rest stop where a car hobo had set up shop. He was enthralled to see us and made a point to show us every inch of his, "Decked out car!" in his own words. Well, his, "Decked out car!" was, in reality, a piece of shit that had no business ever starting again. It was filled with burger king trash and stripped of all seats but the drivers. At the same rest stop, Joey fell in love with a married woman who he actually had the balls to go talk to; respect brother. Later down the line, we found yet another rest stop where a lady decided we were beggars and tried to force $15 on me. I had a pleasant argument with her forever as we debated whether or not it was noble to accept money from strangers. While I hate doing it, I finally took $5 of her $15 to put her mind at ease. At least it bought us buffalo gorgonzola fries at the next town.

Glenwood Springs

O my gosh, we were so hungry! Glenwood Canyon was phenomenal and I never want there to be any more development there, but there was something notable missing from that amazing place… FOOD! That is why we were happy to stumble into the town on the other side of Glenwood Canyon, Glenwood Springs.

I knew I was going to like this when we passed three hot springs before we even made it to the first restaurant in town! Joey wasn't quite as excited about the hot springs, but it seemed Glenwood Springs had something for everyone. In our search for a cheap restaurant to eat at, we passed the 'Doc Holliday Saloon.' Doc Holliday was one of Joey's idols and role models. Known as the fastest gun in the West, Holliday was actually a dentist. Known to have confrontations with many men, and to be the only that walked away when the dust would settle, he quickly gained a reputation for himself. He would later become friends and deputy to Sheriff Wyatt Earp, who essentially hired Holliday as a bounty hunter. Many shootings and close calls later, Holliday would be diagnosed with tuberculosis and sent to Glenwood Springs, Colorado. It is assumed he spent so much time here to bathe in the healing waters, but there was nothing they could do to heal tuberculosis in the late 1800s. Before he drifted off, Doc Holliday said his final words, "This is funny." Supposedly referring to the dozens of times he had been shot at, only to die from an incurable disease instead of a bullet.

We made plans to visit his grave first thing in the morning, but right now we had food to eat. We decided upon the CO Ranch House because it was the closest thing to Doc Holliday's Saloon, which was 21 and up for entry. The greatest thing about going out to eat when you are really hungry is that whatever you get ends up being fantastic! Hunger is the best seasoning for any meal hands down, but it does help that we were eating good food. The $5 the lady had given me earlier that day went towards a massive stack of spicy buffalo gorgonzola fries for us. I remember them being too spicy for any not starving person to eat, but we were starving. My mouth was on fire as I stuffed down more and more fries like a Viking barbarian, #noshame.

Second course came with a chicken Caesar wrap for me and a massive elk burger for Joey. Have you ever eaten something so good that you feel bad eating it because there won't be anymore of it when you're done? That is the look Joey had on his face upon taking the first bite of that burger. Pure bliss and fulfillment for the first bite, then sadness as he put the burger back down. I could see him thinking, 'Why can't I just take you with me? We could have a real small wedding, only about 500 people, honeymoon in the Maldives, sip drinks with umbrellas in them. Doesn't that sound nice?' If that burger could talk, I'm sure it would have agreed with Joey's temporary logic. Unfortunately, that line of thought only lasted a split second before Joey changed his mind and feasted on that burger like it was the last thing he would ever eat.

You would think a meal like that would satisfy us, but we were not done yet. There was an ice cream shop just down the road where we stopped at. We had to enjoy the luxuries while we had them, because they were not here to stay.

We made our way to a little park on the outskirts of town next to some tennis courts and fell asleep listening to the trickles of the nearby stream. For me, that had to be one of the most peaceful nights on the trip. For Joey? Not so much. Joey unfortunately got the downhill side of the tent, which led to me sliding down onto him several times in the night. Unbeknownst to me, I turned into some sort of ninja in the night and kicked Joey right as he was falling asleep several times. He must have the patience of a Shaolin Monk to not kick me back when he later revealed to me that it was far from the first time I had done that to him.

A Day Off

Joey was in a surprisingly good mood for being kicked multiple times the night because we were about to do tourist things for the first time on our trip. First up on the list, go see Doc's grave. It was all the way across town, but Joey took every single step with a smile. Doc's grave lies in a cemetery on top of a hill that is apparently only reserved for a couple local legends, none of which either of us knew. It was only his tall grave in the back corner of the cemetery we were interested in.

It was clearly Doc Holliday's grave before we ever saw the inscription carved into the stone. Multiple decks of cards, full cans of beer, and glass shooters, unshot revolver bullets, and so much cash was strewn across the grave. All memoirs to the things Doc Holliday loved, I guess. I was ready to go in an instant, but this was Joey's idol, we simply couldn't leave after only a moment at his grave. I gave Joey his time alone with the grave just in case he wanted to say some words to honor him. It was funny, by the time Joey was ready to leave, I felt closer to the famous cowboy as well. I made a point to run back to his grave and tip my hat before we left. He really did live an extraordinary life and he deserved respect for that.

Now it was time for my stop, the underground vapor cave hot springs. It happens that Glenwood Springs is home to one of two known natural underground vapor caves in the continental United States. I love hot springs more than just about anything else ever, so you best believe I was going to die before I missed out on that. We walked back across the entire town again to arrive at the vapor caves. I was out of my mind excited when they gave us a brief history on the area.

For centuries, the Ute Indians had visited the vast geothermal marvels of the area that now make up Glenwood Springs. The Ute constructed several mud huts with enclosed doors so they could send the sick in to sweat their demons out. Rightfully so, the vapor caves were regarded as a sacred place of healing for these natives for years, perhaps even centuries before the white man ever came upon the area in 1860. Manifest Destiny and the spread of the railroad took this sacred place from the Ute. Fortunately, even early settlers recognized the value of this place and preserved one of the caves that we were about to visit.

We were led to a small changing room and then escorted down into the depths of the cave. It would be too dark to see, but old-fashioned light bulbs buzzed gently along our descent just bright enough to show the outline of the general shape of the cave. At the bottom of the carved stairs was where our guide left us, we were on our own now. The steam around was filling our lungs with traces of the healing water that now surrounded us. Using the wall as our guide, we ventured even deeper to find two large chambers with primitive stone benches built into the walls. There were people too, I didn't see them at first and jumped when I thought I saw the walls start to move around me. As my eyes adjusted, I could see that they were primarily native peoples, no doubt the lineage of the Ute people that used to hold the rights to these caves. Finding a seat amidst the blackness, I discovered that there were reservoirs of 120-degree water scattered throughout the cave just large enough to dunk your head or feet in. Being that the area seemed far too sacred to speak, Joey and I explored separately. You could feel the history that lived in those caves.

As amazing as the caves were, it was impossible to stay long. We had to take several trips to the surface over our time there. The natives down below could stay for over half an hour in a sitting compared to Joey and I's 10–12 minutes. We took a couple of trips back down into the depths, letting the hot water and steam nurse our beat-up bodies. It was so lovely that I briefly considered buying $120 massage and seaweed wrapping from the women above, but no. We had relaxed for long enough and we were getting restless. It was time to turn north towards Yellowstone.

The Challenge of the Freeway

We tore ourselves away from those magical vapor caves and walked up the street to a gas station to map out the next section of our trip. Glenwood Springs had no northbound roads coming out of it, the nearest highway that would start taking us home was 27 miles west into the town of Rifle. From there, it was another 500 miles north before we reached Yellowstone. We just had to make it to Rifle tonight and that was going to be easy. Our average in the South was around 35 miles, our west average had blown that out of the water so far so that 27 miles wouldn't be an obstacle at all. There was just the issue that those 27 miles had to come from freeway hitchhiking, despite the freeway hitchhiking being strictly illegal. "Hopefully, the cops are cool here," I said.

We found our way up to a turnout on the freeway only to have a man in a huge truck stop and pick us up almost instantly! Add that to my list of reasons to love Colorado. The man driving had to be one of the most generous and open people we met on the trip. Let it be clear that this man was far from being any sort of hippie lover dude though. He was a hard-core conservative and business owner that just so happened to love smoking copious amounts of weed at all times of the day. He informed us that the only time he isn't smoking is when he's driving, which was reassuring. That is, until he told us that he wanted to get a bit of a hotbox going in the car and pointed out some weed that he would be happy for us to smoke. Now, that is an example of about the most generous thing someone could do for a couple of hitchhikers and I really did appreciate the offer, but I hate the smell of weed and do my best to avoid it at all costs. I had done a decent job avoiding weed over the course of our trip, but this was Colorado, aka marijuana heaven. Avoiding weed was much easier said than done here.

I politely declined and told the man I didn't smoke. I could see Joey's eyes light up next to me though, he had discovered weed not long before we left for Charleston and had now gone without it for 18 days. I fully expected him to take the man up on his offer, but after a moment of tense silence, Joey politely declined as well. I waited for our weed smoking driver to drop us off in the town of New Castle before I asked Joey why he didn't smoke with the man. His response made me unbelievably grateful for having Joey rather than anyone else on this trip. He said, "I know you don't like it, Jake." It might not seem like a big thing, but for someone to choose my preferences over smoking something that can make you feel as good as marijuana apparently does? I'll always remember that decision Joey made. It was a sign that he had my back even when we completely disagreed on a topic. That is one of the key things you need in a partner.

We had made a big jump from Glenwood Springs to New Castle, Colorado, but the obstacle of the interstate still stood in our way before getting to Rifle. Deciding to test our luck with the law a little, we wandered out in full view on the side of the highway, hoping to get a ride before the cops saw us first. Hot springs had left us energized; it was time for a show. Joey played some music and I danced my little heart out on the side of that road. I don't think anybody picked us up while dancing the entire trip now that I think back on it, who cares though? It was worth the fun that Joey and I had. I tuckered myself out pretty fast, now it was Joey's turn to step in the ring. The man's not much of a dancer, so I was curious to see what he was going to do. He had a plan though, a 12-feet tall street sign warning of wildlife crossings ahead was conveniently positioned right next to us. Using his rock-climbing skills, Joey climbed up the backside of the sign and started waving to drivers as they went by. The amount of shocked faces was priceless, I even took a picture to commemorate the moment with a picture because it was so incredibly funny. (You can see all our pictures in Joey's Journal starting on page 202.)

All good things must come to an end though. What people don't tell you about that saying, however, is that sometimes good things end because they are replaced with better things. I don't know how we missed him, but I was in the middle of another dance and flipping session when I looked left and discovered a parked cop car maybe 50 feet away from us. "Crap... I guess we saw that one coming." Picking up our bags, we slowly walked towards the cop car, maybe we could make friends?

It has to suck being a cop, you almost never have any idea what you are getting into. It was that same case for this poor guy that got sent out to deal with two random hobos on the side of the road. I can only imagine him pulling his car to a stop on side of the road to watch me doing the robot. Any logical person would see my little show and think, 'O yeah, this guy is on every drug. Meth, heroin, cocaine, you name it, and I bet this guy has done it.' That had to be what this cop was thinking when he saw us. Then we started walking towards him! Some people would have just run us over, thinking they were doing the world a favor by getting us out of it. Not this cop though, he got out and met us halfway to his car.

As Joey and I always do, we gave the officer a firm handshake and introduced ourselves. I could tell he was a little wary of us at first, but when we told our story, all the suspicion went away. We laughed for a couple of minutes and apologized for making him come out and have to deal with us. He assured us it was way better than he expected our encounter to be and offered us a ride back up the exit into town. Are you kidding me? This guy was amazing. Hats off to our officers in the US, they get a bad name because a couple of horrible things have happened in their departments, but at the end of the day, 99.9% of them are all trying to do a good job and make friends while doing it. Although he told us we had to get off the freeway, which was a shame, he was very nice in doing it and by being compliant with him, we even got our second opportunity to ride in a cop car. It doesn't get much better than that.

By the time we got back to town, our officer was talking about how if he didn't have to go take a sergeant's exam, he would give us a ride himself. Just the offer meant a lot to Joey and I. Knowing that someone cared and felt for how hard we were working put a smile on my face. We knew we couldn't expect anything more from the man, he had done as much as he could. We turned to head back towards town, but he stopped us before we could leave. "Hey, I really wish I could do some more for you guys. I respect what you're doing and I want you to make it home. Our department offers us free bus passes every couple of weeks and I never use mine, so take this." He held out a $15-dollar bus pass to us. I couldn't find the proper words to thank him. I just felt elation that someone would go out of their way enough to do something like that for us. He had to leave, so we shook his hand and wished him the best life possible. That is how all police stops should be. With a smile on our face, we walked back into town to find a bus stop. We had our tickets for Rifle.

Spirit Bears

It was lucky for me that Joey pays attention. I got sucked into conversation on the bus again and paid absolutely no attention to where we were going. Joey alerted me when we got to Rifle and let me know that he was getting off, my choice was my own. Scurrying to catch up with him, we found ourselves in the semi-arid desert environment of Western Colorado. This was going to be hard to hitchhike through.

We bought some jalapeno chips and pop-tarts from the local grocery store to support our pure carb diet and looked out across the dry plains that we were about to walk through. The tree we sat under looked to be the last shade we were going to see for a long while. It looked intimidating, but Joey and I both agreed it was time to get back in the grind. We had been spoiled by Colorado, traveling just over 200 miles in less than three days. It had been extremely fun, but both of us knew that it couldn't last.

At last, we stood up from under the safety of our tree and walked out into the hot sun. Although we had hardly walked at all, my legs just refused to comply with my body. We had become so tough since the South, now able to walk 12 mile stretches without any problems with our feet, today was different for some reason. Of course, neither of us said anything, but we could both feel our normal pace was not going to happen today. Once again though, Colorado delivered a wonderful couple that stopped for us. I couldn't believe the luck to once again get a ride when our situation seemed the gloomiest. In fact, I think it was one of my favorite rides of the entire trip due to the surprises that were waiting for us in the mountains.

I can't remember the couple's names at all, but they were some of the sweetest people. They told us all about their kids and what they were doing now that they were all grown up. It felt like we were little kids going to grandma and grandpa's house with them telling stories. Joey and I felt at peace in the back seat; they just gave off vibes that made us feel taken care of. The couple was coming from dinner and not in any hurry to get home, so they asked us, "You two don't mind going on a little safari, do ya?" We looked at each other, I had no idea what they were talking about, but I'd be damned if we turned down a safari!

"Uh, yes, please," Joey and I responded enthusiastically. The husband turned off the highway onto the tiniest dirt road we had been on so far on the trip. We would later discover we were on State Road 252, a place I have filed away in my 'must go back to places.'

At first, our 'safari' looked to only be made up of cow pastures. Not that I was going to say anything, we were happy to just get a ride. As the road twisted and turned, our conversations did the same, passing through politics, religion, and a little flare up when they told us that they don't believe that humans have anything to do with climate change happening. I got a little fired up on that topic, but I bit my tongue and let Joey's cooler mind handle the situation. There is no need to make enemies with people that have been so kind to us. Then we went silent, a small lake had appeared through our driver's side window and suddenly there was a mood shift. The land had changed and now there was a sense of anticipation that filled the car. Our tiny dirt road dipped over a small hill and I could feel that we just dropped off the map…this was true wilderness now.

Birch and Aspen trees enveloped the car and the road turned rough. Our safari guides rolled down our windows and let us listen to the song of the forest. The Aspen leaves rustling in the wind drowned out all other sound, giving me chills up and down my body. The road wound back up a small hill and the valley bottom opened up to show us the huge mountains that stood on either side of us like ancient stone guardians. This was a place forgotten by time, now only the few like our tour guides and us got to witness the breathtaking beauty of it all.

I somewhat remember our drivers asking us questions about our trip, but I was too enamored with our surroundings to properly form a good answer. Joey backed me up and did the talking for me while I focused all my attention out of the left side of the car. When you've spent enough time in the forest, you get a sense of what is around you before any of your other five senses can detect it. That was the case where my attention was now focused, I could feel something was there. There was nothing to see but my laser like focus was not about to waver, there was something that was going to happen very soon.

We had slowed to maybe 10 mph on the roughest section of the road. The potholes bouncing the car up and down as we went. My focus had waned by taking some time to engage in conversation, but my attention was still on the left side of the car. I knew something was out there and I so badly hoped that we hadn't missed our chance to see it… Adrenaline crashed down on me like a tidal wave with no warning. My head snapped back to the left side of the car with force that hurt my neck, but this time there was something there. Two blonde black bears crashed out of the bushes next to us and ran up the hill. Their white backs and brownish paws are burned into my memory despite only seeing them for a second. As they ran, the bushes and trees almost seemed to follow them, allowing the two bears to melt back into the landscape much quicker than anything as brilliantly white as they should have been able to.

Everyone in our car gasped as the bears disappeared. Those are what some people call spirit bears and spotting them is supposed to be an omen of good luck. We all realized how special and exciting that moment was, it was why none of us shut up for the remainder of the car ride about every single animal experience we had from our childhood until now. Babbling the entire way, we wound through a total of 44 miles on that little dirt road through some of the most beautiful land in the country as far as I'm concerned. By the time we saw the next little town of Meeker, it was getting time for us hitchhikers to set up camp. Our safari guides dropped us off next to a campground by the river and drove back up the hill to their mountain home. Paying the $5 fee to set up camp, we went around and explored the little corner of Meeker we now resided in.

We Struck A Deal

Meeker is a tiny ranching town with a population of around 2,300 people. The White River flows along the edge of town, making it a great fishing spot to retreat to for locals and vacationers alike. From what Joey and I gathered, the surrounding hills were where the old and rich retired to like the people that had given us a ride earlier. Our campground was the vacationer's stop, but the great thing about vacationers is that they are the friendliest people you will ever meet.

We set up our tent on the south side of the White River next to the cow pastures, a price we were willing to pay to get far away from the shiny RV's on the north side of the river. We didn't want to deal with drunk people and their dogs, but we still had to re-cross the river to use the bathroom among the vacationers' shiny RVs. We were walking back to our tent after using the restroom with every intention of getting some rest until the most wonderful smell hit our noses. The couple at the nearest RV had just cooked chili dogs, and o what a smell it was. Maybe they saw the look on our faces or they wanted to hear our story, because they refused to let us go to bed until we agreed to sit down and eat with them. Once again, we did not want to appear as needy beggars and tried to refuse, but when someone is offering you food while your belly is growling, you can only say no for so long.

We hit it off with that couple that night by the river. They said they had sons that we reminded them of, so that probably helped our case. It turns out that they were not vacationers as we had thought, they were locals in the process of moving and therefore they were living in the RV for a couple weeks. I don't know how it came about, but we ended up agreeing to help them move the next morning as long as they would pay us with more warm food in the morning. It was the perfect deal and we went to bed with full bellies, a feeling we had come to see as much more of a privilege than the right we thought it was as kids.

Since that night by the river next to the cows, I have come up with a new system for determining how much I like the area we are staying at. I stayed up listening to the gentle wind blowing through the valley, the occasional moo of a cow, the water flowing down the river, but not for the entire night did I hear any sort of artificial noise. No honking, no sirens, not even a pair of headlights interrupted my sleep. It is partially what has made me remember Meeker with such fondness... Annnd a couple of the 10 million+ dollar homes that I wouldn't mind getting in on in a couple years' time! Of the thousand or so little towns that we passed through, Meeker was one of the few places I think I could actually live at.

We woke up to our friends across the river shouting that they had breakfast ready. We disassembled the tent and made our way over to find a bacon and egg breakfast. I looked over at Joey and recited what had become my mantra for the past couple days, "I...love...Colorado, man." I ate as much as they would feed me, knowing that we were going to need those calories soon walking on these hot summer days.

After breakfast, the couple let us hop in the back of their black truck and took us to help them move out of their old home into a new one in the same apartment building. To do that, we simply had to walk down a spiral staircase with a couple bookshelves, two bed frames, and a couch. It wouldn't have been a very difficult task for the Jake and Joey before hitchhiking, but here on day 20, it became apparent that our arms had suffered some losses. While our legs had turned into nothing less than that of tree trunks made of solid muscle, our arms had atrophied. It was a good thing we had been forcing ourselves to do push-ups every day, or that furniture might not have been moved.

Luckily, the difficult pieces were few and far in between. An hour after we started, we moved the last bed frame out of the apartment and claimed our reward of foot long sub sandwiches. Somehow, it came out that we weren't Christian before we walked out, which resulted in the woman giving us the worst pity/holier than thou look that we received over the whole trip. We looked at each other exasperated, 'Here we go,' we said in the silent language we had developed. I let Joey take this one and watched as they briefly argued through a couple different perspectives. Joey's education on physics and mathematics was too much for the woman to comprehend, so in response she stopped answering his questions and just started stating what she thought should be true. If she wasn't willing to have a conversation, there was nothing we could do but listen as she forced her prayers upon us, "Dear Lord, please show yourself to these young men before they get home. They are going to need you." Let it be clear, I have no problem with anyone being religious, but when someone puts themselves on a pedestal above us because we are 'lost souls' or whatever they want to call us, it pisses me off big time. A shame to end what had been such a lovely relationship on a sour note. No matter, we were free from Meeker, now it was our goal to cross the next state line into Wyoming.

The Angel of the Prairie

Walking out of Meeker was supposed to be fun and exciting, but walking out onto the highway, we were met with little to be hopeful about. A vast prairie stretched out in front of us. No trees, no shade, and no town for the next 21 miles. We thought our walk in Tennessee was hard from Chattanooga to the town where we met Santa Claus, but in Tennessee it had been wet and we had shade; here there were no breaks unless we wanted to roast in the rising heat of the sun. I froze for a second in the face of adversity, but Joey and I's unsaid rule of not complaining to each other held true. We didn't know how, but we were going to make it through this.

What I kept forgetting is that we were still in Colorado! The only state in the United States that actually likes hitchhikers. We were maybe a mile into our walk when a blue van drove by us a little slower than normal, but didn't stop. Wasn't the first time that had happened, no need to pay it any attention. Then the same blue van drove back by us on the other side of the road, I caught a glimpse of the girl inside. She was studying us intently, for what I wasn't yet sure, but it was good news that someone was thinking about picking us up. The van disappeared for a bit after the second drive by. A couple of minutes later though, the same van drove by us yet again and pulled to a stop just down the road. I told Joey that the van ahead of us had been watching us for a couple of minutes already and to walk slow. We didn't want to scare a potential ride off, but there was no need for my precautions. The girl that hopped out of that car was fearless.

She was much younger than I had expected, a freshman in college and a fantastic artist, Katelyn earned her name of 'The Angel of the Prairie' the moment she hopped out of the car and approached us. Her long blonde hair and Aladdin pants wisped in the wind as she walked with a smile stretched across her face. Joey and I both took our hats off to show our faces and introduced ourselves. She was thrilled to hear the brief version of our story, with a wave of her hand she turned and motioned us into the car with an air of authority that left us no choice but to follow her.

Katelyn Krehbiel was her name and she definitely has a spot among the top three most memorable people on the trip. Every time I thought I was getting a feel for her, she would throw in another random fact from her past that would completely throw me off. Her list of hobbies was too long to count on both my hands, including paragliding, slacklining, highlining, backpacking, river raft guiding, rock climbing, dancing, singing, and art. Katelyn's story is a bit of what you would consider most people's dream life. She had taken trips with her friends all over the northwest to explore almost all of the area's natural wonders. Not to mention the solo trips she takes all the time through some of the most remote wildernesses available.

Her charm did not end there though. The longer we spent with her, the more attractive her personality became. The carefree and nonjudgmental attitude shone through from the beginning, but what I think it was that made me feel so comfortable around her was the yes attitude. So many people find it so easy to say no to taking a risk, going out on a limb, trying something new, having fun. We all get that little twinge of fear when we are getting ready to attempt something new, and for most people, it stops us in our tracks. 'If it's scary, it must not be worth doing' seems to be the consensus of thought. Katelyn is the total opposite. For her, fear was nothing more than a sign of the right path to take. Not only was it very admirable, it was inspiring.

We spent just under 300 miles driving through Wyoming with our Angel of the Prairie. She is one of the few people that we met on the trip that I think we might see again. I think there is still a lot to learn from her young, yet wise and pure soul.

Donna the Biker

Katelyn dropped us off at Super one in Lander, Wyoming (the first grocery store we recognized). Funny, the little things we appreciate. From there, Joey insisted upon finding a place to do laundry. I was reluctant but eventually caved to his requests. I hate laundromats, I didn't want to do laundry, I didn't like paying to wait 90 minutes, and I felt that it was futile if we weren't also going to take a shower. We had already gone 300 miles today though, it's not like we were in a hurry.

I didn't go in with Joey upon arriving. Instead, I decided to relax outside and try to nap a little. The minutes ticked by and I grew impatient, *What was taking him so long?* I finally followed Joey inside to see what we were waiting on and suddenly became so grateful Joey had insisted on this laundromat. There, sitting next to Joey, was the most incredible woman I have met to date.

Her name was Donna, and she is also found on my top three most memorable people list alongside Katelyn.

While impossible to place Donna precisely on that list, she was definitely the survivor of my top three, giving off a tomb raider-esk vibe. Donna doesn't own a home, she doesn't have a husband or children, and had spent most of the last six months alone, bicycling back and forth across the country. What… A… Badass. She needed a job though, right? Just so happens that she worked for a company called Backroads where she took people on bicycle tours all over, not just the United States, but the world. She speaks like four different languages and showed us pictures of her leading groups in Chile, Argentina, Italy, and a number of other places I can't recall.

When she asked about our trip, we told her of all the amazing places we had been only to have her respond, "O yeah! I've taken five or six tours there." She had been everywhere we had been so far at least twice. We didn't have a candle to hold against that beautiful…amazing…pretty…strong…okay, I was a *little* in love with her. Alright, maybe a lot, dream woman hands down actually. Unfortunately, she was in her 30s and more of a solo player. What a shame… Joey patted me on the back as we watched her ride off into the sunset, "Tough luck, bud." Joey really could read me like a book.

Music Festival?

Traveling 300 miles for the day had satisfied us, it was time to enjoy the little town as much as we could. Music in the distance drew us to a large park that happened to be celebrating a town music festival where camping was free. Joey bought us some mac and cheese, I bought some pulled pork, and we sat to enjoy the last bit of the music festival.

Sitting there on the grass listening to music with good food was one of a handful of brotherhood moments that I've had with my close friends, I never forget nights like that. We even found a nearby baseball field and tested out Joey's old pitching arm. It was still real good, much to his pleasure and much to my pain. My hands hurt for the first time on the trip from trying to catch Joey's fastball, awful idea. The day drifted away to night. We bonded over our hate for a chihuahua that wouldn't stop barking and drifted off.

We slept in until maybe 8 a.m. feeling lazy. All the sitting we had done the day before had thrown our bodies for a loop. Regardless of how we felt, there was a day to attack here. Our treasure searching zone was within 300 miles now, the same distance we had traveled yesterday. We were getting excited, but before we got there, we still had to get through Teton and Yellowstone National Park, which meant plenty of adventure before treasure hunting.

It Takes All Kinds

I was entertained by Lander while we were walking out. It gave me a similar feel to Sandpoint, ID where Joey and I were from, except for the fact that it didn't have a lake. The town was a little dry for my taste, but it did remind me that we were starting to get very close to home.

Not a mile outside of the city limits, we were picked up again. I was loving this luxury treatment and this ride was particularly important. Wyoming is a real desert for 80 miles or so between Lander and the lush national parks in the north. It would have been miserable to walk through. Thank the heavens that was not something we had to worry about with our new ride; a married couple heading to their new piece of property in Dubois, about 80 miles north of Lander.

They were the kind of nice people that I enjoyed for the most part, but probably wouldn't trust them with anything. The husband was a lawyer for those who could not afford a lawyer. You know the part when the cop says, 'If you cannot provide a lawyer, one will be provided for you'? Yeah, that was this guy that would be provided to protect the scum of the earth. It was almost sad to hear how good this guy was at his job too. He told us about one of the more memorable cases he had gotten where a man had stolen some women's clothing to try it on in the bathroom. I don't agree with the stealing, but try on whatever you want, right? It got bad when a woman walked in on the man halfway through dressing himself. According to our driver, the man 'panicked' and slit her throat with a knife. O, you see, the story doesn't end there though. She had a baby with her, he slit the babies throat as well and tried to run for it.

Me: So, you ended up defending him?

The Lawyer: Yes. I got him off with an insanity plea and a possibility for parole.

Joey and I: …Hm.

To be honest, I was livid. I couldn't even understand how the cops didn't shoot the man in his cell after they found out what he did. The man driving us had enabled that psychopath to have a possibility to do it again! Next time, it could be his wife that walked into the bathroom when a bastard dressed like a woman slits her throat. I give the lawyer credit; he had defended dozens of poor people before that who couldn't afford a lawyer. That being said, enabling one of the monsters of society to have a chance at freedom is something should keep him sleepless most every night. What really got to me at the end of our car ride was that if this lawyer had managed to defend people like that once or twice, it meant that damn near every town in the US had a lawyer that had done the exact same thing. Now think of all the monsters in their cages just waiting for their time to see the light again…

"It's a Vagina, Not a Clown Car!"

Our lawyer and his wife dropped us off in Dubois at a gas station upon request. We were hungry and the landscape was starting to turn back into forests. It was a place I'd like to see again without a doubt. I grabbed my usual pop-tarts and granola bars, Joey got his usual can of Chef Boyardee and some chips, it was a normal day in the life of hitchhiking until this freaking guy pulled into that gas station parking lot in his RV.

He walked by us stumbling a little with a gash in the back of his head. "Oooo…good thing we're not getting a ride with that guy," I told Joey. I had to open my big mouth. We were taking shifts outside with the bags while the other went inside to use the restroom. It was my turn on shift and I saw the man with the gash in his head start eyeing me. *O no,* I thought. *If he offers us a ride, we can't say no, Teton is still 55 miles away and we need to get there one way or another.* Part of me wished that he wouldn't give us a ride because of how goofy the guy seemed to be.

That is exactly what happened though, Joey and I didn't take two steps off the sidewalk before the man shouted out, "Hey! Where are you two heading?" One thing led to another and we were sitting on the couch in his RV waiting to get a move on.

Joey and I were used to making the first move in a conversation on these drives, most people just aren't all that good at talking. This man was not one of those people, in fact he was so far the other way that we didn't have to say a single word for him to tell us every single story and experience he had. It started with the story about how he got the gash in his head, "Yeah, man, I got drunk and high for the first time in a couple years last night with a buddy and ended up falling down." He pointed to some broken glass where it appeared that a head had smashed into it. "My head hurts pretty bad now, how's it look to you guys?" We looked at the dried blood plastered to the back of his buzz cut and back at each other.

A consensus was made, "You're fine! You got a little cut, I guess…but it's not that bad…kinda."

The man smiled, "Great! LET'S GO TO TETON!!!" he shouted in his best singing voice. We had no idea what the next 55 miles held with this crazy dude who introduced himself as Donny Lewis.

We started the drive curious about his life, so he started listing off the most insane feats of any human ever. First was, "Hey, hey, hey! You see that mountain over there? That mountain is named after my great grandfather. He was one of the first pioneers in this area, and the guy was a badass. You know I'm a badass too, right?" Then he would look at me with his mouth half open waiting for me to laugh before he got too tired of waiting and would laugh plenty loud enough for all of us.

Next accomplishment was that he was a national champion high school pole vaulter. "O really?" I asked him.

He replied, "Yeah… Probably would have gone to the Olympics if the Oakland Raiders hadn't recruited me to play in the NFL with them." This guy was so funny. He continued to tell us that he played linebacker and quarterback for them before he was then drafted into the NBA with the Golden State Warriors where he played from '87–'89. It all sounds like the biggest lie you've ever heard until you listen to all the detail that he went into about every experience that he had. He was naming off players like Karl Malone in basketball being his friends and other stories that didn't sound made up. It almost made me want to believe him!

Other things he had done according to him were:

1. Founded Diablo Solar Services, the number one solar company in the world.
2. Survived avalanches in his pastime.
3. His brother was one of the key pieces in creating the entire internet and invented the scoring system for fantasy football.
4. Graduated with two bachelor degrees in archeology and some sort of chemistry.
5. Was fired from working in Antarctica for becoming too close with a penguin.
6. Had a personal role in integrating people at Guantanamo Bay.
7. Was a master construction worker that had started his business in Japan, eventually making enough money to travel all over Asia, having sex with women the entire way, of course.
8. Had met multiple celebrities like Tom Cruise and a couple of other obscure names I didn't know.
9. His mother had invested $180,000 in (Microsoft, I think) back in the early '90s. Every time he wanted to make a point, he would reach into his wallet and grab his card saying, "You have no idea how much money is on this card!"

I almost peed myself listening to this guy talk. Suddenly on a serious note, he turned to me and said, "I have to show you guys something." Without any time for Joey and I to argue, our driver veered us off the road in the massive RV onto a tiny dirt road into the forest. Well, this meant one of two things, either he was going to kill us, or we were about to see the coolest thing of our life. I made sure to quietly slip my knife into my pocket just in case things got crazy when we stopped at a campground two miles from the highway. Donny stumbled out of the driver's side door, probably still hung over from the night before, and let out his two wiener dogs to come on a walk with us.

The comedy would not end with this man, watching him walk with two little wiener dogs on his heels was about the funniest image I could have conjured up in my mind, but this was real! We walked about a mile out into the woods, Donny talking about how many times he made some cheerleader cum back in college the whole way, to see a waterfall. It was very pretty, so I turned to him and said, "Thank you for bringing us out here. This is awesome."

He looked at me like I was the most uneducated piece of shit he had ever seen and said, "That's not what I brought you out here for," as if it was obvious. He kept walking down the trail to a tree that had been cut down in the shape of a chair. On the seat of the chair were initials in a heart, "Yep!" Donny declared. "This is where me and my wife had sex for the first time." Joey and I couldn't hold it in, we laughed with Donny for ages out there in the forest. This goofy dude had driven us all the way out here, to then walk even further, to show us where him and his wife had sex for the first time. You have got to be kidding me.

The fit of laughter passed and one of us asked him, "Are you still together with your wife?"

"Nah, I got rid of that bitch years ago," he said and started on his way back to the RV. My god, this guy was something else.

The next stretch of our ride focused mainly on sex. Donny talked about how he fucked his best friend's cheerleader girlfriend back in college, he talked about his most recent woman and how he hated having to sneak around her kids, even gave out free how to make a girl orgasm tips. He had even given each of his moves their own names like, "The Fishook, The Brooklyn," and kept saying, "Next time you're in bed with a girl, try this!" None of his tactics sounded particularly pleasant for either party.

Our time was prolonged with him when Donny started complaining about his pounding head from the hangover he had given himself the night before. "We have to stop and take care of this," he said. Any normal person would assume that he meant to take a nap or maybe another walk in the woods, it's what I thought as well until we pulled into the Red Fox Saloon. Donny assured us he was a local legend in these parts and that he could get us in to have a couple beers with him. We briefly considered ditching Donny and letting him raid the bar by himself, but to be honest, it was too funny to not go watch what was going to happen inside.

We followed our stumbling driver inside the bar and as he promised, there were no ID checks so far away from town. The first thing Donny did was perhaps the funniest thing he did on the entire trip. A redheaded waitress walked by us, prompting Donny to turn around and whisper excitedly, "O my god! Did you two just see that redhead? I love redheads!" Turning back around, Donny approached the waitress and asked, "Hey, sweetie, are you from Tennessee?" Joey and I put our hands on our faces, we knew exactly where this was going.

She stopped and looked at Donny, "No, I'm from Croatia." Completely unphased by this new turn of events, Donny finished out his line like a champ.

"Cause you're the only 10 I see!" he proclaimed as if expecting to get kissed right there for being clever enough to rehearse the dreaded Tennessee pick-up line.

Joey and I were dying from laughter behind him, but it only seemed to encourage Donny to continue where he left off. Motioning us to follow him, we approached a married couple. The wife was beautiful and maybe in her early 40s sitting next to her handsome husband. That didn't phase Donny though. Stumbling through the bar right up to the edge of their table, Donny reached deeper into his pickup line repertoire for something more sophisticated. "Excuse me, ma'am," he said, completely ignoring the husband. "Were your parents thieves?"

Bewildered that someone would accuse her parents of such an act she retorted, "Well, no, of course not!" Donny smiled like a snake; she had just fallen into whatever trap it was he laid for her.

He replied in an equally shocked voice to the woman, "Well, they must have been! They stole all the stars out of the sky and put them in your eyes."

Joey and I were having laughing fits that were leaving us breathless behind Donny until that point. He might have gone one step too far with that one, especially because the husband was right next to Donny's new target. We were a little excited, Donny might get us in a bar fight and that would be an awesome story to tell! The wife giggled a little out of sympathy for how horrible Donny's line was and looked to her husband to see what he would do. He looked at his wife, back at Donny for a second, and smiled while shaking his head. Chuckling out of pure amusement, the husband patted his wife on the shoulder and stood up to use the bathroom. Joey and I wouldn't have laughed if it weren't for the confused look on the wife's face. Her mouth was wide open in shock, but we could tell she was trying hard not to laugh at the pickle she found herself in. I don't think Joey or I laughed harder for the entire trip than that moment right there. I don't remember how we finally got Donny to leave the poor woman alone, but it didn't take long before he was throwing out another line to the bartender. "Hey, are those galactic pants? Because your ass is out of this world!" You can imagine the reaction that one got.

Donny had two beers before we convinced him to leave, assuring him that he didn't want to be drunk driving. He agreed and walked out the door to us, all the while telling us, "Man, did you see that blonde chick? It's like she was looking right through me to a bed!" Delusional too? C'mon, Donny. The last stretch of the trip to Teton was filled with Donny telling us he knew the name of every bear in the state and that they come to him when he calls their name. We even stopped a few times for him to shout out into the woods, "Susie! Come here, Susie! I have some friends that want to meet you!" Despite all this crazy behavior of Donny, I really don't think he was actually crazy. He was just a guy that loved to have fun in any way he could. More so, he was completely free from the judgement of other people and there was a lot to respect about that.

Stopping a couple hundred feet short of the check in station, Donny grabbed my arm and lowered his voice for the first time since I met him. "Now, you listen to me, son... *pause for dramatic effect* It's a vagina, not a clown car." Busting up laughing after a couple more seconds of silence, Donny waved his goodbyes and left us on the side of the road to contemplate his last words.

"It's a vagina not a clown car," I recited to Joey. What do you think that means?

"Never have kids," Joey said with a smile on his face. He was right, I think that is exactly what Donny meant.

Teton National Park

Donny had dropped us off in one of the most beautiful places I had ever seen. The Tetons shining in the distance almost brought me to my knees. This was the place we needed to be. There were only a couple minutes for that until we went into mountain man mode. According to Donny, the entry fee into the park was about $70. Ninety percent of the time, I would be so happy to pay that to support our national parks, but this was not a normal time. The Greyhound ride had drained my bank account a lot, we didn't know how much longer our limited funds were going to last. That was reason enough to justify sneaking into the park for us.

Creeping along the river side, we infiltrated the park and found a nice log to sit to watch the river flow. A river rafting crew came floating by, the guides shouting out orders to eager guests doing their best to stay afloat. The sun shining on the Tetons gave them an almost purple hue. Everything came together in that moment as perhaps the most at peace I have ever been at.

I scarcely remember the rest of the day other than we didn't get a ride, the Tetons had given me presence. There was nothing else to focus on other than the beautiful mountains and forest around me, so I lost myself in it. What I do remember is the beautiful field that we chose to sleep in. We made a bear hang and set up the tent right before the mosquito swarms settled in on us. So many that they actually blotted out some of the fading light of the sunset by blanketing our tent. Their little needle mouths reaching for our blood made me so happy that we had brought a tent on this trip. Being the smart ass he is, Joey said, "I think we should let them in." That caused me to lose my mind for a few seconds until I realized he was joking, and we settled in for a long night's sleep…or so we thought.

We had gone to bed around 7:30 p.m. to prepare for our long day of walking tomorrow. About three hours later, we were woken up by a flashlight shining through the holes in the tent. Snapping awake, I let my mouth flow and shouted, "We good, you good?" to whoever was outside.

They responded "Ummmm, yeah. Please step out of the tent." Being very careful to show our hands first when crawling out into the darkness, we stumbled out unable to see anything because of the still bright lights in our faces. They were Park Rangers here to flag us for not sleeping in a proper campsite. When we told them our story, they seemed slightly more empathetic, but not enough for us to be able to sleep the rest of the night. They told us a campsite was only a couple miles up the road and that we better get there if we were going to sleep again. With that, they drove off and left us alone in bear country. Awesome.

What choice did we have but to walk? There we were in the middle of the night, walking along the side of the road. Of course, the rangers were too busy to give us a ride, why bother driving us five miles up the same direction as you are going, right? For a couple hours, we walked through the dark with not a single person stopping to ask if we were okay, if we needed a ride. It just seems like common sense to me when you see someone that could be potentially hurt or in trouble, you stop for them. Not the case in Teton, I guess.

We got a couple laughs about how uncaring the rangers seemed to be at first, but eventually we were steaming at the fact that we had to walk through the dark because no one would give us a ride. We decided that there was no way we were going to find a campground in the night, so we turned off on a side road and found a gas station. We thought that the bears wouldn't be coming this close to a building at night, so we fell asleep on the hard concrete next to the gas station.

What do ya know, 3 a.m. rolls around and we are woken by yet another flashlight in our face. I have no idea who this guy was, but he never asked if we were okay or needed help, the first words out of his mouth were orders to leave the gas station. There are no cities around here, no ghetto, no anything harmful. How often do people in Teton see hitchhikers? I would bet money that almost none of them had ever seen anybody in our situation before, yet they still assumed that we were looking for trouble. The man's flashlight beam never wavered as it tracked Joey and I packing up, yet again, and walking off into the dark.

We made it about 300 feet away before we stopped caring if we camped in the woods and got attacked by bears. We found our way behind the gas station sign at the beginning of the driveway and decided it was the best medium between no bears getting curious enough to check us out, and no dirtbags with flashlights being able to find us. It worked well enough that we slept for the next two ½ hours until the sun finally woke us up with dirt and pine needles clinging to the sides of our faces where we had laid our heads.

Thanks, Idaho!

It is good to remember that Joey and I were still repping the cardboard signs on our backs. Something we had seen the night before was that there were a lot of Idaho license plates driving past us. "We might be able to play off of that," we decided. We walked back to the gas station at 6 a.m. and threw away our old signs that said, 'Have a great day!' and dumpster dived for some fresh cardboard. On these signs, we wrote, 'From 7B Idaho!' and 'Hitchhiking back to 7B!' Both signs were supposed to target our Idaho family and let them know we weren't crazy people. We thought it was a great idea and walked back out onto the road expecting to be picked up instantly.

It was still too early for vacationers it seemed, the roads were devoid of any sort of traffic. I was actually glad for the opportunity to be able to walk some more today. We had been so spoiled over the past couple days that we hadn't hardly had to use our legs. Now, we got to enjoy the morning in Teton on foot, a way that few people get to see it.

About 8 a.m., the traffic picked up, first it was California plates, then Utah, then finally at around 9 a.m. we saw the first Idaho plate wiz past us. While they didn't stop, it gave us hope that they were here. As the day ticked on, Joey and I must have seen about 50 Idaho plates zip by, even a couple from our hometown of Sandpoint, indicated by the 7B on their license plate. Even though we might have gone to school with their kids, they might have been our neighbors, or maybe even knew them personally, not a single one of them stopped. There was no love from our Idaho family despite our 7B signs.

Pam and Bob

We walked for several more miles along the beautiful winding road until we had to stop at one of the more daunting signs that we had seen. A flashing traffic light read, 'Mother bear and cubs ahead, use caution.' Given that Joey and I were on foot made us pretty nervous. The last thing we wanted to do was piss off a momma bear. In light of that, we spent about 30 minutes waiting for someone to pick us up right in front of the sign, hoping that it would further encourage someone to pick us up. It didn't seem like Teton was a fan of the hitchhiking type though. We had only one choice to keep walking, waiting had proven to be a waste of time almost every time we had tried it. Even if we got attacked by a bear, it would be a badass way to go out.

We didn't see any bears, but we did see one of my favorite creatures in the world called bison. We passed a whole herd and listened to their snuffing and stomping with elation. We knew full well bison are just as dangerous as bears, but for some reason they just seem so much more chill with people. I thought back to Kansas and how magnificent it would have been to see the mighty bison roaming across the plains there 500 years ago. Unfortunately, now the great beasts are almost exclusively tied to the Yellowstone area if you want to see them in the wild.

About seven miles in, we made it to the next rest stop with food and restrooms available to use. Teton had come to life; tourists were bustling about left and right in their fancy RV's and trailers. I could just imagine them going home and telling their friends how they went camping out in the remote wilderness of Wyoming. It was hilarious to think about while Joey and I sat on a picnic bench absolutely filthy from sleeping in the dirt the night before. I felt smug and happy knowing that we had been camping for real without a tent the night before while all of the rich tourists pretended to be campers. Joey and I were the real deal.

Joey had gone to the bathroom which left me alone to quietly giggle to myself watching the tourists that were so out of their element. Watching intently, I caught the eye of a woman that I could tell had just read my, 'Trying to get back to 7B!' sign on the table next to me. I took an instant liking to her and her husband as they briefly debated on whether or not they should pick us up under my watchful eye. The way they talked to each other just looked right.

Both had to be from 40–50 years old. They were a clean looking couple unlike some married couples I see that don't particularly care about what they look like around each other. They looked great and their body language spoke of high class and high-status individuals. They were a novelty in the crowd of tourists in that they were the only ones that seemed calm and composed among all the bustling cars coming in and out of the parking lot. Still watching under the brim of my hat, I saw the man turn around and look at me. Nodding at his wife, he put a brilliant rich man's smile on his face and started my way. I smiled, knowing that we were about to get a ride.

The man introduced himself with a strong voice as Bob and offered us a ride north to Yellowstone. I happily accepted and told him I would meet him over there just as soon as my friend came back. With a smile, he assured me that they would wait a few extra minutes and strode back to the cross between a van and RV that the woman waited in. As soon as Joey came out, I gave him the low down which ended up in us running across the parking lot to our new friends. I shook Bob's hand and met Pam, the woman that had originally seen me.

We did our best to be extremely respectful and keep our distance knowing we probably smelled awful, but Pam especially seemed so curious about us. Joey and I moved up closer to them and ended up having a fantastic time over the next 40 miles. It seemed Bob was one of the richest men I had ever met up to that point. He owned a string of 12 or so restaurants from Louisiana to Florida that he had started himself. He shared that up until very recently, he had been married and had kids that Pam was only partially acquainted with. *Ah, this was starting to make sense,* I thought to myself. Pam then told a little bit of her story, that she was also recently separated from her husband and had kids of her own. Something about just the way she talked made her seem like one of the most attractive women that Joey and I had met so far, but under her beautiful exterior was a caring soul and an iron that left no mistaking that she was a woman to be respected. Bob was the successful entrepreneur and Pam was the gorgeous woman that every man hopes to find himself with one day. Both recently separated? I'll leave the conclusion up to you.

I assume that they were probably all the way out in Wyoming escaping the trouble their relationship had caused back home. While, initially, I was a little shocked at what their situation seemed to be, I was so comforted by the fact that they were together. It represented a self-serving nature that I think is totally lacking in most people. I ask people all the time who the most important person in their life is. They rattle off answers like my mom, my brother, my best friend, and what they don't seem to realize is that they are all wrong. It is YOU. You are the most important person in your life and Pam and Bob represented that perfectly. They were out here on the road serving themselves and living their best life doing so. I don't care who they left behind, these were people that deserved the ability to pick up and leave whenever they chose as far as I was concerned.

The most of our drive consisted of me trying to extract whatever business advice I could out of Bob and Joey flirting with the beautiful Pam. It was up there with the best rides we had and I am very thankful I got to meet those two amazing people that have been so much of an inspiration for how I live my life. We took a picture with them when they dropped us off at West Thumb in Yellowstone. We turned our backs to leave but heard them shout from behind us. Holding up their finger telling us to hold on, Bob ran back inside the RV and brought back out two tinfoil wraps and a huge bag of huckleberries. He told us, "We were fishing on the lake yesterday and caught some fish that we cooked up Southern style. If you eat them with the huckleberries, they are amazing! Take them." We tried to refuse but the offer was too good to pass up. With a smile from those two brilliant people, they drove off down the road and left us with our bounty.

Cultural Differences

Jogging down a short little path to a rest stop, Joey and I gorged on the mini feast that Pam and Bob had left us with. We were starting to realize how much of a luxury it is to be full. The greasy fish and delicious berries were going to keep us warm in this new and much colder environment.

West Thumb is a rest stop on the west side of Yellowstone Lake. The walk to the lake involves boardwalks that take you past geysers and hot pools with strange plant and bacterial life that live in them. Those pools are part of what makes Yellowstone so famous. As we stood to go see the pools for ourselves, a massive tour bus pulled into the parking lot and unloaded what seemed like over 100 people. What I found most bizarre about it though was that they were all Asians.

Let me be clear that I do not have anything against Asians, but I just found it bizarre that an entire bus would be filled with nothing but one ethnicity. While I am not racist, I will admit that I was no longer particularly excited about visiting the boardwalk because of the behavioral differences between us and the bus full of people. I have never been to any part of Asia so I don't know firsthand what about our cultures are different. That being said, the people that got off that bus behaved differently than I have ever seen anybody do and I did not like it.

The first thing that I instantly registered as different about them was that they were completely absorbed in their own world. I am not sure I would act much different if I was in a foreign country with a bunch of my own people surrounding me. Nonetheless, the first thing I saw one of the people off the bus do was bounce off an American man's shoulder. There was no apology, not even a registering of what just happened; the Asian man just continued toward the boardwalk still engrossed in his conversation with his family.

Because Joey and I are social butterflies, we couldn't pass up the opportunity to have a new experience and followed them down the boardwalk hoping to strike up some sort of decent conversation about the amazing journey it would have taken to be here in Yellowstone from Asia. Then something I've never seen anyone do happened right in front of us. A family stopped and knelt down to grab something out of their bag. This action caused the trail to be temporarily blocked, so Joey and I stopped respectfully and waited for them to finish whatever it was they needed. Apparently, the Asian people behind us didn't see the path as blocked though. I was shocked when a whole family squeezed by me, gently pushing me out of their way and sandwiching Joey and I against the guardrail as they proceeded to step right through the family that had stopped in front of us. I would have taken that as disrespect if I didn't see the family stopped in the middle of the path in front handle it like it was an everyday occurrence. None of them even flinched as they were stepped over and around by the people behind them.

Initially, I felt ready to push back when someone pushed me out of their way, but when the two Asian families passed in front of me, I realized I was looking at a cultural difference between America and places like China and Japan. I swallowed my urge to retaliate and patiently waited as people continued to push by Joey and I despite them never apologizing or even making eye contact to give us some sort of human connection.

Finally, the family in front of us got their act together and pulled out an electronic device I can only describe as a periscope. It looked like a larger than average tablet that had periscope handles attached to the sides of it. As we followed the family, I realized that not one of the families in front of us ever put down their electronics to simply admire the beauty of what was around us. They all just looked through their screens so they could press the buttons on the side of their periscope to take a picture.

It was horrifying to me. These people were living in a completely different world than Joey and I had been living in for the past 21 days. There was absolutely no presence about these people as we walked around the boardwalk. So many of them had these crazy periscope looking things that they never stopped looking through, it was one of the most insane things I have ever seen hands down. The whole time we were there, not one of them said a word to us. I even had a kid bump into me so I laughed and caught him to make sure he didn't fall. "Be careful, pal," I told him with a smile. Instead of a smile and a thank you back, I saw fear etch over the kid's face. I let him go immediately and watched him run back to his parents and point in my direction, speaking what sounded like some form of Mandarin. The parents looked at me and huddled around their child, quickly walking the other direction. That was one step too far for me. These people were acting unlike anyone I had ever met and I wanted nothing else to do with them.

Looking back on that experience, I really do realize that we were experiencing a cultural dynamic that we had not been exposed to before. Joey and I were coming from this present, suffer-through-it kind of mindset while the tourists with their periscopes were from a life of comfort in the city and had never seen anything as beautiful as where we were in Yellowstone. I won't pretend to understand it, but it's probably important to recognize them as different rather than bad.

A Man's Man

I will say that there is one thing I adore about hitchhiking. There is no way to predict what will happen next at all. It seemed whenever Joey and I were dying from walking for miles and miles, there was at least another 10 miles to walk. Whenever we were fully prepared to walk those 10 miles, there was a ride right around the next corner. That is what happened walking out of West Thumb in Yellowstone, not even a mile in before we got another ride from a couple that would tell me one of my favorite stories from our trip.

I don't remember what they looked like or what their names were, what I do remember is their presence. Both the man and the woman were around the 70–80 range. Most people would hear that and think they were old, that is not the case. They were children in grandparents' bodies that had never lost their sense of adventure.

They picked us up in a silver rental car and were happy to make room for us to hop in. I remember asking if they were traveling with anyone else, but they told us that they were very much alone. I was so impressed that an old couple like themselves was taking their own vacation to go see the world. They weren't retiring in a mansion in Arizona, they were embracing their curiosity and still exploring. I loved that about them and for them to include us in that was such a high honor.

Before they took us out of the park, they warned us we had to stop at Old Faithful. The husband had been here over 30 years ago and had to see the place with his own eyes again. Walking inside the Yellowstone Lodge, I felt like Joey and I were their grandkids along for the ride with how warmly they treated us. They told us stories about how the lodge used to be and how the park had changed over time. Unfortunately, we didn't end up staying long enough to see the geyser erupt. Everyone seemed to be ready to get a move on.

My favorite part about our time with those people was on the last stretch of driving together when we asked them how they met. They smiled at each other and told us a story I will not soon forget. They had been boyfriend and girlfriend for a few months before they were separated due to unfortunate circumstances and forced to break up. There were no promises of getting back together or even certainty that they would see each other again. Lucky for them, life seems to have a way of putting you in the right spot at exactly the right time that you need to be there. After many months apart, the man returned to her hometown and was shocked to find her still there, but with a boyfriend. That little tidbit of information would have crushed most guys. They would think something like, 'She has a boyfriend? I can't intrude on them.' Then they would fade away. His approach, however? For the sake of reenactment, let's call him Tim and her Martha and show you how the interaction went below.

Tim had heard Martha had found herself a new boyfriend since he left. Though he hadn't met the fellow yet, he heard from the whole neighborhood how great of a match they were for each other. How they laughed walking around town together without a care in the world... 'Like we used to do,' he thought.

She had to know he was back just like he knew of her and her new boyfriend. 'Would she break up with him for me? What is going to happen when I see her again?' All these thoughts ran through his head until he saw them for himself sitting at a table in a dinner together laughing. There was a slight twinge of pain in his gut, but just looking at her again was all he needed to know that she was the woman he wanted to be with for the rest of his life. Walking inside the dinner and right up to the edge of the table, he took a minute of silence and looked at the boyfriend to analyze every bit of him. The silence grew heavy as he looked back to his future wife and asked,

Tim: Is this your boyfriend, ma'am?

Martha: Well, uh, yes.

Tim: Are you going to marry him?

Martha: Umm, I... I don't know.

Tim: That's what I thought.

Without another second's hesitation, Tim took Martha's hand and they walked out of that dinner together. Today, over 40 years later, they were still holding hands and telling their story to a couple of hitchhikers from Idaho.

That story hit me hard. Not only was it a beautiful love story that defies the statistics of divorce, it also showcased what it is to be a masculine man that goes after what he wants. You see, our generation of men in the teens and 20s have been castrated by society. As a whole, we are not manly, we are not outspoken, we don't protect and fight for what we believe in! No, instead we sit on the couch and play video games and our voices break when we try to ask out that beautiful girl down the street; what happened?

That is the question the wimp version of me was asking myself riding in the back of that couple's car. It's not that the Jake that sat there wasn't strong physically, clearly evident by my hitchhiking across the country with my brother, it's that he was weak when it came to social interactions and influence. He let other's reality define his even when the situation had nothing to do with them. The man in front of me had created his own reality his entire life and chosen to live with his dream woman forever because he made a decision that **he** was the alpha male.

I have learned so much more about what it is to be a strong masculine figure since then, and if there is one thing I can share with any of you, it is this. There is a war on masculinity today. Terms like toxic masculinity and movements like radical feminism revolve around the idea that the only way to be strong is somehow to castrate us, the young men and boys of society, to become mindless drones that don't stand up for anything.

REMEMBER THIS: When you become a strong alpha male, it is not taking away from <u>anyone.</u> It is only raising up those around you to become the best versions of themselves.

While I can't go into this topic fully in a book about hitchhiking, I will say that the realization that I was giving to others by amassing more power and influence rather than taking has changed my entire life... Next book topic?

Tim and Martha dropped us off in West Yellowstone and left us one last important message. With a power in his voice I had not yet heard, Tim viciously shook both of our hands and said, "If anyone ever tells you can't do anything! You look them in the eye and tell them to eat a bucket of shit." You could see the pain in his eyes as he spoke. I did not know our trip had affected him in such a way but it did. We were an inspiration to him that the next generation still had adventurers and future leaders. He shook our hand a final time and returned to the warm embrace of his wife. They waved out the window as they drove by us and back into Yellowstone.

Something changed in me watching their car drive away. It was a life changing moment that didn't change me then, but showed me that there was another way to be rather than a soft and malleable little boy. Looking at myself now, I can finally say that I have made some of those changes. And like we all should, I am still getting better every day.

Detective Joey

The cold had come for us in west Yellowstone. The wind and high elevation were too much for us to tough through anymore. We had to stop at a gift shop and buy some warmer clothes before we could make it through the freezing cold night in the local park. We made it through the night hardly sleeping due to the frigid wind, the only thing keeping us going was the possibility that tomorrow could be the day that we find Forrest Fenn's treasure. That fact alone kept us smiling every time the wind woke us up with an extra cold gust. Now, looking back on that night, I see it for what it was, an omen of very hard times ahead.

Waking up the next morning was absolutely thrilling, we bought a local map and sat down to analyze all our options and where the treasure would be located exactly. It was a good thing I had Joey with me, he is a detective in the making. I tried to help for a couple of minutes but I had not the slightest clue on where to go next. I did my best to help, but eventually, Joey looked over at me and icily said, "I love you, bud, but you need to stop sighing and be quiet." At that point, I was more than happy to sit back and watch Joey work his magic.

Now that I was out of the picture, I could see Joey start to fire on all gears. With the poem in one hand and a map in the other, Joey narrowed his eyes and went to work with a sharpie. Not 30 minutes later, he had the area narrowed down to a couple of peaks and a valley that seemed like the closest match to the poem. Still to this day, I have never seen anybody dial in their inner Sherlock Holmes as much as Joey did in that moment. He has a detective's mind, and that mind was about to lead us into the unknown. Grinning from ear to ear, we walked out of West Yellowstone confident that we were walking towards a very rich payday.

Fortune
Finding the Treasure

The Countdown

It had to come to an end at some point. We could only get so many rides before the good luck ran out. Walking out of West Yellowstone was that day that the good luck left us. I had been waiting for this moment and was almost excited that it had finally come. What a fitting test for ourselves to have no rides while being so close to our destination.

As we were hiking up a steep hill about six or so miles in, a pickup truck stopped for us. He gave us a ride for about three miles and gave us some oranges which was very kind, but the best part was that he was from one of my rival schools growing up! I had probably faced off against him on the football field as a freshman in high school. Just another sign of how close we were getting to home.

We waved goodbye to him and looked up the long stretch of road he had just dropped us off at. I couldn't help but laugh, in the South we could count on their being a gas station at least every 15 miles or so. Here in Montana, we were no longer given that luxury. The road we stood on wound around a lake and over the crest of a hill in the distance. It was such a scenic view that was about to test how bad we wanted that treasure.

As if Fate had deemed we did not have it hard enough walking so far, the wind started to blow. Not some gentle breeze that ruffles your hair, this wind was the ice-cold kind that chills you to the bone no matter what you are wearing. Given that Joey and I didn't have that many clothes to start with, we were freezing. The only way to keep warm was to keep up as fast of a pace we could. That meant we didn't get the luxury of stopping very often over the remainder of that day. The positive part of that is that you realize how long a single day is and what is possible to accomplish in the 12–16 hours that we were awake. Standing on top of the hill on the other end of the lake that we had seen so many hours earlier was a marvelous feeling of accomplishment.

The downside to walking that far is that first, we killed our legs. That wasn't the really bad part though because that amount of aching pain kept us present. It's almost fun to fight against your own body when it is telling you to stop. The really bad part was second, we were burning an insane amount of calories walking that far. By this point, we were used to going without food for most of the day, but we would usually make up for it with a big meal of extra granola bars at night. That big meal was no longer an option when we finally stopped by the river on day 22. I had not properly prepared enough food from West Yellowstone, which left my body with only one option, start eating itself.

Going to bed hungry is not ideal, but the gentle trickling of the Madison River next to us reassured me that everything was going to be alright the next day. "We have to find somewhere with food soon, right?" I remember saying to Joey. Probably feeling the same hunger pains I was, he nodded without making eye contact. For the sake of saving energy, we didn't say much else for the rest of the night. If that had been our first day on the trail of the treasure, we were going to need everything we had to get through the mountains.

The start of day 23 was a gorgeous one. We woke around 6 a.m. with the first ray of sun and saw that, not only had our tent frosted over, but the outsides of our sleeping bags as well. I put a hand on my stomach to gage how hard my body was working. I could barely even grab the skin of the lean muscle that had developed over the past few weeks. It would have been great if I was planning on going to the beach in the next couple days, but it was far from ideal for what we were about to do. The cold outside forced us to hide inside our sleeping bags for another 45 minutes before we could get out and deconstruct the tent for another full day of walking.

The hunger was starting to get to us as Joey and I became short with each other. It's very hard to think about anyone other than yourself when your belly growls at you every couple minutes. I had to rely on the natural beauty The Madison River to save my mood while my stomach lectured me about not eating enough…

Just as my hunger was starting to get real, we ran into a campground with food and WI-FI for us to check our location and get filled up on clif bars. $2 a bar was taxing on our limited budget, so I made one of the worst decisions on the entire trip. After I used the restroom, I came back to find Joey with four clif bar wrappers scattered across our tiny round table in the campground grocery store and busily looking at maps on his phone. As I downed five clif bars, I heard great news from Joey, "It looks like there is one more place with food before we head out into the mountains." That was perfect! That meant I could do minimal shopping here and stock up at the next store before we go out for the next couple of days. I ate another clif bar and bought four more for extras before we left that lovely RV park next to the river.

With full bellies, Joey and I were a little more at ease with one another for the next couple of hours. We passed the time by doing our best to psychoanalyze each other's personalities and why we do what it is we do. It's amazing you can know someone for so long and still not know so much about them (look for more details on this conversation in Joey's Journal). No rides came our way over the next 13 miles as we walked. Before long, all my extra clif bars were gone. "I'll restock as soon as I get to the store," I kept telling myself. Well, the store on the horizon finally did appear, but food was not what awaited me there.

I thought I was walking into salvation when I walked into that fly-fishing shop; I was wrong. It was a beautiful store with the most gorgeous homemade Damascus steel knives I had ever seen. Fishing rods and fish food lined every counter and shelf, but there was not even a granola bar in sight. That meant either Google was wrong, or Joey was wrong and honestly, I didn't even want to know which. I quietly stepped back outside and sat down on the porch bench. The mountains we were about to tackle loomed in front of me, snow still marking the top of the tallest ones. I could see the cold nights that lay ahead of us while traversing those mountains, it was not going to be easy. Dumping my bag out, I did a quick food assessment, half a loaf of bread, three granola bars, and a single Babe Ruth chocolate bar. I ate one more granola bar right there and packed up all my supplies before Joey came back out.

I had figured that it might be smart to go for the next town before we mounted an assault on those intimidating 10,000ft mountains. I had not taken into account how much this meant to Joey though, he walked out those doors with an aura of excitement and anticipation about him. I could only smile at him, there was no use in being mad that the store didn't have food, it's just what was. I brought it up to him right there, "We got a decision to make, Joey. We can either walk off into those mountains tonight, or try for the next town and come back for them."

Before I had even finished speaking, Joey had made his decision, "No, I think we should go now. Can you make it?" I looked down at my bag and up to the huge mountains again. I knew full well that if we went into those mountains, I was going to starve on the meager food supplies I had brought.

Taking a quick moment of silence, I looked back at Joey and asked, "You still have those freeze-dried meals, right?" He nodded. Breaking eye contact again, I looked back into those snowy mountains. Even the extra freeze-dried meals weren't going to be enough for that. I slowly looked back at Joey and nodded, "We said we wanted to suffer out here, right?" Joey smiled his Joey grin and helped me up off the bench. The decision was made and the countdown was over, it was time to tackle the mountains.

The Start of the End

For Joey, the whole trip was leading up to this moment. I hadn't seen him so happy since the train ride. Crossing the highway was crossing a threshold. For over three weeks, we had walked on the side of the road and never been far from asphalt. Now we were leaving out into the open wilderness.

Just like the day we first started, the butterflies stirred in my stomach like when we were about to take our first steps on the journey. At the start of the trip, I had taken a moment outside of our hotel to find the little bits of beauty around me. The sun shining through the thick foliage is the one I distinctly remember. Probably because it was a sight I knew I could die thinking of. Now that we were starting a new section of the trip, it was only right to stop for a minute once again to take in what could be our last days alive.

This time when I looked around me, I found a very different world. The green snowcapped mountains that could be seen in every direction showed so much more nobility and power than the city of Charleston. The horses that ran across the field in the distance showed the freedom we had given ourselves by separating from the warm and cozy world we had been raised in. Lastly, the lonely highway that we had just stepped off represented the fact that we might never step foot back into the modern world. We both knew that we might just die in those mountains to any number of things. That was once only a scary thought, but now it was more. Sure, the fear was still there, but now there was a realization within it. That is what life is supposed to be. Not safe, not knowing that you will drive back home down the same road that you have every night, thinking of the warm bed waiting for you. Life is supposed to be scary and uncertain; it is the only way you finally look up and realize the gift of being able to breathe in another gulp of air. That moment right there standing at the edge of the mountains is what I would consider the happiest moment of life because of the uncertainty facing us.

Let's Go Look for Treasure

Walking back into the deep forests that Gallatin had to offer was such a breath of fresh air. We had a tiny taste of the wild in Glenwood Canyon and Hanging Lake, but we were always surrounded by people. Now, we were totally alone in the wild. It was nice, but it most certainly came with its own set of challenges.

By the time we found a nice meadow to camp in, we had covered over 20 miles in one day yet again. Such a trek two days in a row was well deserved of a campfire. We collected some soaking wet wood together and somehow managed to get a fire started with our lighters. Part of me felt at home, but there was a feeling of dread I couldn't shake. Maybe it was just the shortage of food, or maybe it was my sixth sense kicking in. I just knew that whatever was coming was going to be harder and different from anything we had dealt with so far. In light of that, I left Joey and walked around the next bend in the trail with my phone to make a farewell video.

I would hope that a video like that would be played at my funeral so I composed myself. I took a few minutes to center myself with some meditation and organize my thoughts. When I was satisfied with my state, I flipped on my recording device and made a short video. I watched it over many times before I went back to Joey. It started with me talking about the friends' funerals I had already been to. Haden Kistler and Derek Lowery were two very close friends of mine that I had lost in high school, one to cancer and the other to suicide. I told the imaginary people watching on the other side of the screen that I didn't want my funeral to be like theirs. I wanted everyone to know that I was okay with death, I was at peace and fully aware that I might not come out alive and that it was a conscious, thoughtful decision.

After watching it for about the fifth time, I stopped and thought about how sad and accepting I sounded in my video. Why would I be so under enthused dying doing what I love to do? I deleted the video and remade the whole thing with fire pouring off my tongue. The one thing I remember saying was, "Hey, just so you can all know and be at peace with it. Joey and I died fighting for every bit of life we could hold onto. We never gave up, so do us a favor and never give up for us, will ya? Thanks." I felt far more at ease with that video and more energized about the next portion of our hike. I looked up at the huge mountain fading in the dark and said out loud, "You better bring it on, because we are bringing everything we got." Some might say it is only a fool that challenges mother nature, but I look at whatever punishment she can dish out as a gift that I got to witness just some of her immense power. Let me tell you, folks, she took my challenge and ran with it.

The fire was still roaring when I got back to camp. Joey sat by the light of the fire, journaling just like he had been when I left. "How was it?" he asked.

I smiled, "As good as I could ask for, man." He looked up from under the brim of his hat and smiled back at me. There was still such happiness about him. Without saying anything more, I sat down and pulled my half loaf of bread out of my bag. The only thing I had eaten was the 10 cliff bars I had bought at the RV park earlier. It was a lot more than I had the day before, but nowhere near enough to nourish my sore muscles that had walked 40 miles over the past two days. I had a choice to make, I could ration my food and be weak for however long we were going to be out here, or I could eat my fill tonight and face the effects of starvation on our way out. I chose to eat my fill and finished off my loaf of bread and a granola bar so I only had one left. Those calories would at least get me through the tough hike tomorrow without any complaints. I just tried not to think about the hike down.

Joey woke on day 24 with an infectious excitement about him that rubbed off on me. He was so thrilled to have a whole day for us to look for treasure. I shared most of his excitement, but to be honest, I was way more excited about the hike. I viewed the treasure as an added bonus. If we found it? Amazing! If we didn't? I was mostly out there to see the mountain lakes at the top anyway. I calmly looked from side to side for anything that seemed out of the ordinary or matched Finn's clues. I was already starting to feel a rumble in my stomach, reminding me that I needed to save as many calories as possible.

Joey did not feel the same way. When we got into what we considered our, 'treasure hunting territory,' Joey was keen to check every possible hiding spot. I was happy for him; he was born to be some sort of searcher like a detective or a treasure hunter. I did not participate in the early searching though. I was going to save myself for the most likely spots and not waste my energy on the many. So, we walked, much slower than usual, I'll say, because of Joey's searching, but I didn't mind…yet.

As we started our accent, we found ourselves on a rock bluff with what seemed like infinite possible hiding places for a 40–50lb chest. This was a place I would happily search with Joey. Leaving our bags behind, we descended down either side of the smooth stone crafted by the glaciers so many years ago. There were ledges and animals' trails, even a tiny cave that seemed like the perfect hiding spot, but no treasure. I did one circle around the bluff that took about 40 minutes and returned to my bag. I had searched enough to be satisfied that the treasure was not here. I expected Joey would be getting to the same conclusion soon so I waited…and waited…and waited. I couldn't hear any tumbling rocks from Joey's maneuvering no matter how hard I listened. That either meant that Joey had gone so far away that he was out of ear shot, or he was not moving. The latter option waned on my mind as I listened to the quiet forest around me.

I knew Joey was fully capable, but accidents happen. I started shouting for him, "Joey! Joooeeeyyyy!" No response… *Ah shit,* I thought to myself. I didn't want to assume anything, but I needed to know he was okay. I did another short lap around the bluff that yielded no results. My phone didn't have service and it would take a minimum of a full day to find a place that did have phone service to call someone for help. I calmed myself and thought back to the river rafting trip we had taken for our senior graduation. Joey had flown up and down that mountain like he was born there, he had to be able to handle these cliffs. I played my last card and sat down to wait.

I was right, about 15 minutes later, Joey walked out of the woods. I could see a slight bit of annoyance on his face when he discovered I had only searched around the bluff that we sat on. As I pulled out my last granola bar, I thought about ways I could sneakily ask Joey why this meant so much to him despite his claims that he doesn't care about money. Out of the two of us, I am far more willing to say that I want a massive amount of money as a reward for bringing great services to the world. Joey on the other hand? He hates the idea of money and was proud to say it since the start of the trip.

I thought that if I simply asked him, 'Why do you want the treasure so bad?' It would only start an argument I didn't have the energy to deal with. Instead, I started what I hoped would seem like a friendly conversation by asking him, "What would you do with your share of the treasure?" He seemed mildly interested in my question, so I kept pulling on the string and tried to figure out his real desire underneath the dollar amount of money. At first, I thought it was status to flaunt around his school when he talked about taking care of his friends, second I thought it might be power when he talked about all the wrongs in the world that he wanted to fix. I sifted through the information until I found what I was looking for in him talking about the next trip we would take together. Joey wanted FREEDOM.

It is the exact same thing that I want from money. The freedom to go when and where I choose, the freedom to support something you believe in, etc. It really was a sneaky and manipulative way to suck information out of Joey, but that was not the big mistake I made. I think he realized what I was doing when he stopped and said something like, "Jake, do you really want to find treasure?" It's probably not what he said at all, but it was the question he was really asking under the surface. It was there that I told Joey the closest thing to a lie I've ever said to him. "Of course, man," I responded. "We came out here to find it, right?" The translation for that would have been something like, 'I mean, it would be cool to find it, I guess. I'm really just hiking this mountain because of how bad I know you want to find it. I would much rather go camp out in one of the cool valley bottoms on our way out of the forest and maybe come back another day to search for it.' Looking back on that talk, I probably would have solved a lot of upcoming issues if I had just told him that I didn't really care if we found the treasure or not. Instead, we stood up and continued up the mountain.

Have you ever noticed that you can sort of understand someone's intentions no matter what comes out of their mouth? Yeah, Joey knew I wasn't being completely honest with him from the beginning. He just really wanted to believe that we were out there for the same reasons; TO FIND THE TREASURE! That was not the case and Joey was starting to realize it. I knew Joey was one of the most logical and intellectual people I had ever met, but I was counting on a lack of social intelligence that would allow me to let him live out his dream. Yeah, good luck. Joey knew exactly what was going on, he just loved me so much that he started lying to himself because of what I said. I'd like to say that me insisting we should go find the treasure to cater to Joey's wants came from a good spot, but it really just came from the shame of not wanting to let my partner down. That mistake cost us big.

A Trip Back to Primal Man

I really tried hard over the next hour to look for treasure with Joey, but I just didn't have the energy to pretend like I thought the treasure would be hiding on the side of the trail. If it was going to be anywhere, it was going to be at the mountain lakes that we needed to get to before dark. Joey checking under every rock was not helping us get anywhere. Hours ticked by slowly as I watched Joey hop off the trail every couple hundred feet and search a possible treasure hiding spot. It didn't take long before I could feel the tension in the air. We both knew that the other was getting mad, but there was no energy to address the problems on an empty stomach. It would end poorly if we brought it up while continuing to hike into the forest.

Just as I felt we were about to explode, we stumbled upon a place we both agreed could be a good hiding spot for treasure. We had started walking along the side of a stream that ran parallel to the longest ridge line I have ever seen. Joey and I agreed that the ridge above us sounded similar to a place Forrest had mentioned in his book. Maybe we were onto something

We stopped and took a moment to explore a big rock slide with numerous well-hidden away spots. I searched for 30 minutes or so and didn't find anything. To me, that meant, "Let's go see if we can find another good spot to look." To Joey, that meant that we needed to triple our efforts and scan this rock slide from top to bottom. *God damn, this is going to be a long couple of days in the woods,* I thought to myself. I sat down and let Joey search to his heart's desire while I thought about how we were going to keep this brotherhood together.

I was already extremely annoyed with Joey and it was only our second day in the forest. I knew my annoyance came mostly from my empty stomach that had burned through my food from yesterday like a furnace swallowing coal. I did my best to stay logical and calm as I waited. Deep breaths and meditation kept my hunger at bay, and once again increased my tolerance with Joey. Little did I know, Joey's form of managing his anger was to just keep searching. As long as he had a task in front of him, he could stay cool. That was another problem, we had totally different coping mechanisms as to not explode on each other that just kept adding to the wave of anger that was soon to come crashing down.

When I was younger, I got involved with the one of the widest spread basketball training programs in the world known as Elite Guard Training. I bring this up because of a foundation principle that they taught in the program that I have since adopted as one of my own. The key to basketball, just like the key of any major sport or skill-based trial, was to practice enough that you could take your mind out of the equation and allow the instincts you have programmed into yourself to take over. It was so hard for me to even begin to grasp this concept until I was presented with a metaphor that likened our mindset and emotions to waves crashing on the beach.

If you have never visited a beach, it is a magical experience to watch the waves crash with such power so tirelessly. I would recommend taking your kids as soon as possible so they can learn this philosophy. I was a child when I first saw the ocean and was taken back by awe in seeing the waves. My curiosity compelled me to get closer to the water and understand this strange phenomenon of water that could stand up and move. My awe was quickly replaced with shock when a wave rose up and smacked me in the face with enough force to knock me down. Every survival instinct you can think of went off as I was tossed back onto the beach by the force of the ocean. 'This is the end,' was all my little mind could think as my mom scooped me up and dusted some sand out of my mouth with a chuckle. It took two more smacks in the face by waves before I learned a very important lesson. You can't hold back a wave.

With emotions and our mental health, it is a similar concept. If you let them flow, the emotion will lap peacefully on the white sand beaches. Fight it and the tide rises against you until it builds up to something you can't hold back, eventually crashing down in the form of a tsunami. It is a great metaphor for what was happening to us, the wave was getting bigger and soon, there would be no holding it back.

Joey didn't seem like he was going to finish looking any time soon, so I started slowly walking up the trail without him in an attempt to get him to follow. It kind of worked, but Joey was not a fan of being herded. Miles passed painstakingly slow as we made our way higher up into the mountains. That was far from the worst part though, the hunger pains were starting to come in waves for me. Not eating was one thing, but forcing yourself to keep hiking higher and higher into the mountains was another thing entirely. I had to factor in that we still had a whole day's hike out as well, something that Joey seemed totally oblivious of.

By late afternoon, we reached a beautiful mountain meadow that was perfect for camping. It was a small meadow that had two flat levels to it. The trail came in on the uphill side and dropped down to become level with the stream flowing past us. We had walked under a thick canopy for most of the day, but now the sun shone in all its glory on us. Refusing Nature's urges to keep walking in the sun, Joey and I found a nice camping spot that would let us hear the stream gurgle by when we went to bed. After the tent was up, I took a moment to explore as to keep my mind off the growing hunger pains.

The ridge still ran across the horizon on our left, telling the story of ages ago when the glaciers carved these mountain valleys out. This ridge was different in the fact that it was a massive granite wall that boxed us in on our left that not even Joey, with all his rock-climbing prowess, would be able to climb. Across the stream on our right loomed the intimidating snowcapped peak of the mountain we were trying to climb. It was an absolutely gorgeous place that was still very wild. Walking down to the stream, I found that nature had included a bathtub for our stay in the forest. Sticks running down stream had gotten caught in the perfect fashion that acted as a barrier to the fast-flowing stream. This left calm waters in the makeshift bathtub that were perfect for a soak other than the fact it was freezing cold. While it wasn't the greatest soaking temperature, mountain streams have some of the best drinking water in the world. By drinking straight out of the stream, you get all the good-for-you minerals and bacteria that the recycled water of the big cities denies you.

I could only enjoy myself for so long before Joey came searching for me to go higher up the mountain to see if we could find anything. Reluctantly, I submitted to Joey's will and got ready to go. The good part was that we got to leave our bags behind with the tent this time. Not having those 40lb bags on our back made life infinitely easier, allowing us to jog as we continued up the narrow trail. With my bag off, I was more than willing to help Joey search now. A cave and another rock slide made for great searching zones where we both enjoyed ourselves a little bit. The tension between us was easing, everything seemed to be swinging back in the right direction until we heard those rocks fall.

At first, we thought it was a little rock slide, but we quickly realized we were listening to foot falls as something large making its way down the steep mountain. I froze and gripped onto my bear spray just in case. The rocks kept falling, whatever we were listening to was on the other side of some thick vegetation and moving very slow. I had the high ground over Joey so I whispered just loud enough for Joey to hear, "Joey, on me." I had every bit of seriousness in my tone that should have told Joey to get his ass over next to me so we could make the best decision on what to do together.

My eyes were still locked on the thick vegetation where the noises were coming from, but I caught Joey glance up at me out of my peripheral vision. Instead of hustling up to meet me, however, this motherfucker adopted the most lazy, carefree body language he had on the entire trip and looked back down to keep searching crevices for the treasure. I was about to yell at him, but the footsteps were closer than ever, and the echoing of the bowl-shaped part of the mountain we were in made it impossible to tell how close they were. My brief anger at Joey not grasping the seriousness of the situation was replaced with survival instinct when I saw a flash of brown fur through the bushes. I hadn't even seen its face, but there was no mistaking that I had just seen my first wild grizzly bear.

Suddenly, a rock fell behind me. I snapped around expecting to see another bear, but instead I saw my infuriatingly dumbass partner still searching through the rocks behind me with no care in the world. I whispered as intensely as I could, "WHAT…are you DOING?"

Clearly annoyed, he sassily replied, "I'm looking for the treasure." I caught myself before I exploded on him, I didn't have any time to deal with his bullshit. Looking back into the bushes, I caught a final glimpse of the fur before it disappeared into the next section of vegetation. Though I could no longer see anything, the footsteps of the creature were still very audible. I listened to them go for a couple of minutes before I slowly turned around to look at my partner who clearly had no care for either of our safety. I have never physically shook from being angry, but looking at Joey in that moment, I was livid. I was about to start screaming, but lucky for him I heard a particularly loud foot fall from the bear once again further down the mountain. It wasn't worth it to get mad and have any uninvited guests come to check us out.

Bottling up everything that I wanted to unleash, I sat down and took a deep breath. There was not a chance in hell that I was going to help Joey look for anything at that moment. I spoke out loud, "When you're done, we need to go back." I didn't have to look at him to know he was pissed too. This was getting to be absolutely ridiculous for me, he didn't seem to comprehend that the real goal at stake here was that we still needed to get out ALIVE.

We walked back down the mountain cautiously together, scarcely saying a word to each other the whole way while we tried to deal with our own personal anger. It didn't take long to reach the tent where Joey confronted me and told me that it didn't seem like I wanted to find the treasure as bad as he did. I was completely open now, "No, I am way more concerned with the fact we don't have any food and have no idea how long it's going to take us to get out of here."

A look as if he had been betrayed crossed his face, "I thought that was the whole reason we came on this trip?" he said in a low and controlled voice. I came an inch away from punching him in the face right there. For over three weeks, he had vehemently told me how much he hates the idea of money. Now he had the audacity to say that the only reason we came on this trip was to find gold? *Fuck you,* I thought to myself. The only way I could make sense of it to myself was to think that the hunger was affecting him more than me. Maybe if we got some food in us, we would feel better?

Joey still had two freeze dried meals left, that would have to be enough for the next couple days. We started a fire and Joey started a breakfast egg meal on his camp stove. I searched my bag a last time to make sure I didn't have any food leftover, but the only thing I found was a single Babe Ruth candy bar. I tucked it back in my bag as a last resort and went back out to sit with Joey. I was very adamant that we needed to leave soon. The hike out was going to be a lot harder than our hike in (we had chosen a route before we entered the forest that would bring us closer to the next town and was much longer than our hike in). Joey did not want to leave though, "Why do you want to leave so bad, Jake?" He asked me questions without ever making eye contact with me now. In his mind, I had completely betrayed him.

Slowly, I responded, "Joey, we don't have enough food to keep doing this." He looked up from under the brim of his hat and looked me in the eye this time.

In the same low and purposeful tone, he said, "You should have come prepared."

My blood went cold, *Did this piece of shit really just say that to me?* I stopped everything I was doing and tried to stop myself from leaping across the campfire to show him exactly how I felt about him in that moment. *Was he not the one who told me there was going to be food at the fly-fishing shop when there was none? Did he not listen when I had told him that I didn't have much food to go out into the mountains? Was he not the one that convinced me to hike out here with him anyway despite the fact he knew I was low on food?* Of course, I couldn't say that though. Silently, I stood up and walked away from the campfire out into the darkening forest to cope with my anger.

About 15 minutes later, I walked back into the campground and found Joey sitting there with a full pan of eggs that was untouched. Joey started to say something to me, but I held up my finger before he could finish, "Not…yet," was all I could say. There is nothing I would have liked more than to eat those eggs. My mouth watered at the smell, but I was far too angry to accept food from the guy I was ready to kill right now. I walked into the forest again and composed my thoughts so I could tell Joey exactly how I feel.

I punched a couple trees and kicked some logs trying to think of a way I could let all this anger out and not ruin our friendship. He was still my brother after all and I wanted to finish the trip with him, but I just felt like he was starting to make such bad decisions now that we were so close to the treasure. We had talked numerous times about how important it was going to be not to get treasure fever and had always agreed that each other was more important than any tangible thing we could find. He had agreed with me every time we brought that topic up and now what was he doing? Getting treasure fever, exactly what we had said we should not do. Composing exactly what I needed to say, I marched back to camp ready to fight.

Joey was still sitting in the same spot patiently waiting for the argument he knew was coming. I took a deep breath and started, "Joey, I have tried to be very flexible with whatever we do on this trip."

Interrupting, Joey interjected, "Me too." With a sigh, I caught the other words waiting on my tongue. Joey still refused to make eye contact with me, instead electing to stare into the fire. I couldn't be mad at him; we had been through too much together for me to hold onto so many negative emotions about him. A huge weight was lifted off my shoulders as I let my eyes drift to the fire as well. Don't get me wrong, I still thought he was being a moron clinging onto the idea of finding this treasure so tightly, but now at least I wasn't mad at him for it.

My anger now deflated, I lowered myself to the ground next to the fire. "Okay, Joey," I said, "I'll help you look for it tomorrow." What I left out was that if he didn't come to his senses by tomorrow, I was going to have to either leave him or force him out of the woods. My body was sending every warning signal that I needed food to my brain, headaches, growling stomach, extreme fatigue were all starting to hit me. We needed to get out soon or pay a hefty price.

Joey nodded, looking at me for a moment this time with a little surprise in my change in mood. We each ate half of the eggs Joey had made from his freeze-dried package, it should have made me feel a lot better, but if anything it was just a tease to my empty stomach. Half the meal couldn't have been more than 400 calories. Just enough to wake my metabolism up for me to realize how hungry I was again. We both went to bed hungry, our growling stomachs seeming to take turns telling us how stupid we were for being out here without any food. I definitely file the experience of "sleeping hungry" away into the 'I never want to do this again' category. It took a long time, but eventually I did fall asleep.

Nightmares woke me up early the next morning, probably my body's way of telling me to get up because I need food. My stomach felt like it was collapsing in on itself, a sign that there was no more time for sleep, we needed to get out of here ASAP. We gathered the supplies for a quick fire and cooked Joey's last freeze-dried meal. The 400 calories didn't do much, but it was way better than nothing. Leaving our bags once again, we started up the steep trail that we had gone up the night before.

My hunger still ate at me, but we both held true to the no complaining rule we had set in the beginning of the trip and kept going. It took all of my energy to hold a steady pace up past where I had seen the bear the night before. The rising sun helped warm us and the beautiful scenery helped me lose partial feeling for my growling stomach. 'This is good for me,' I kept telling myself over and over again. We had wanted to suffer and now we were getting our first real doses of real pain.

Surprisingly quick, we arrived at our destination. It was a large bowl-like valley not quite at the top of the mountain that allowed for the formation of small lakes and insulated the winter snow from being melted. I wanted to see all of the natural beauty and split off from Joey to marvel at the spectacular old trees that dotted the alpine environment. Despite all my entertainment with the whitebark pine trees, I had made a promise to Joey the night before that I would help him find the treasure. Tearing myself away from the beautiful trees, I turned and realized I had lost Joey. I wasn't particularly concerned given that there were only a few lakes that he would be searching around, it would be no problem to find him.

Slowly walking around the edge of another lake, I found a patch of snow that preserved the tracks of the wildlife that frequented the area. Speaking out loud and humming, I catalogued the prints. "Lots of deer, some elk, ooo that might be a little pika, and... shit." If there was any doubt about what I had seen the day before with the brown fur, all the proof I needed lay right in front of me now in the form of a couple massive bear tracks. Putting my hand down next to it to compare size, I guessed that the track was about 13–14 inches long. The long straight claws at the end confirmed that this was a grizzly bear. "Hope I find Joey before the bear does," I muttered to myself.

It wasn't easy, but I did eventually find Joey exploring a wetland area. The water channels were diverted by a thick, bright green moss that was tolerant of living mostly submerged by the water. It reminded me of an African wetland when the rains come back on a much smaller scale. Hopping from rock to rock across the little wetland, I made it to Joey who was busily searching the edge of the rockslide. I tried a few attempts at conversation, but he made it clear that he was not interested in talking. I retreated to a safe distance where I could keep an eye on him in case Mr. Bear decided to come check us out and admired the amazing place we were in.

I could see the peak of the mountain rising above us. The north side facing me was still covered in snow. It was not one of those things that gets less scary when you get closer to it, if anything, it made me really respect how hard this hike had been. Climbing to the top of the mountain looked like suicide from where I was sitting. The snow covered what I knew was another rock slide of flat, shale-like rocks, there was no visible trail, and the wind was blowing hard enough that I could faintly hear the whistling blowing across the mountain top. I have so much respect for nature, it was humbling to look upon so many beautiful sights in one day.

A growl from my stomach interrupted my sightseeing. The hunger had briefly faded after our small breakfast, but now it was back in full force. Sighing, I reached into my pocket and took out the last scrap of food that I had, the Babe Ruth candy bar. I took my time unpeeling the plastic wrapper and made sure to tuck it away deep in my pockets so I would not accidentally litter in this amazing environment. I held my prize up to the sun to look up and down every curve of the extra 200 calories in my hands. Babe Ruth was going to be the last boost of energy to carry me out of Gallatin National Forest. Taking my sweet time, I gratefully nibbled my way through that candy bar. My stomach thanked me by easing the growling for the next 30 minutes while Joey continued to search. I laid down and took off my shirt to bathe in the sun. I wanted to enjoy the feeling of having something in my stomach for as long as it would last.

In the meantime, Joey had finished searching and caught a glimpse of me sunbathing on rock. While I'm sure he did his best to remain calm, I know that seeing me laying down infuriated him. Trying to manage his anger is probably what took him so long to finish searching the mini wetland. Composing himself, he walked over to me to offer his next idea. I heard him coming and sat up to scan the lake, a half assed attempt to look busy on my part. I looked at Joey and asked, "Think we should head back down?" Silence filled the air as his gaze drifted up to the peak of the mountain.

Pointing, he said, "I think we should check there."

I was taken back in complete disbelief. I had been scanning every inch of that peak and already declared it a suicide mission even if we had food! Now he wanted to try to go up to the top wearing some shorts and a flannel with no food? "No way man," I told him. There was no way hiking up that mountain ended well for either of us. Even if we found the treasure, it's not like we were going to eat gold. No, we needed to get down immediately. More silence filled the air while Joey pondered my words.

"Okay," he said. Without a second word or even a glance in my direction, he started a brisk pace back towards the trail we had come up on.

I felt initially happy that we were finally agreeing to go down, but guilt quickly filled me. Even if it was the dumbest idea ever, I should let him do it if it's what he really wanted to do. Running to catch up to him, I stopped Joey and offered him my bear spray. "I don't want you to resent me for not letting you go," I told him. The look on his face told me that what I had just said was the last thing he wanted to hear. Breaking off our talk, Joey started running down the mountain away from me. I took a step to chase after him, but thought better of it. He needed his time to be alone.

It had been two nights since I had challenged Mother Nature's wrath by saying, "Bring it on," when I looked up at the peak from the base of the mountain. Symbolically, she decided to answer my wishes at the same time Joey and I were at our worst, emphasizing that our partnership was over. Looking up to the long ridge that sat to the east of me, I saw a large storm cloud coming our way. Thunder echoed and lightning flashed within the black clouds to let me know that I was about to get every bit of my request from two days ago. I couldn't help but smile as the rain started to fall. There was great humor in this bad situation. Taking a minute to enjoy the cold rain, I started down the trail towards camp alone.

Not five minutes passed before the storm was right on top of me, flinging lightning bolts through the sky and hailing balls of ice large enough to sting even through my shirt. It got bad enough that I had to stop under a tree and take cover from how hard the hail was falling more than once. Then a rumbling sound louder than any thunder should be echoed through the valley and up to me on the mountain. The rumbling turned into a deafening crashing that was unlike anything I had ever heard, almost like the planet was groaning in pain from the lightning that was striking it. I was so deep in the trees that I couldn't see what caused it for sure, but the only thing that could have been that loud and continue for that long was if one of the granite walls on the ridge had broken off and fallen to the valley floor. This was insane and put even the storms in the South to shame. We were at the mercy of Mother Nature now, hopefully she would be merciful enough to let us exit her domain.

It took a long time to get back to camp in the storm. On the way in, we had been wondering how the trails were so well maintained, but now it became clear, they were dry river beds from storms like this. There was a constant water rushing around my ankles that was getting high and higher the longer I stayed out in the storm. This was going to be a long trip back.

I found Joey taking cover under a tree upon arriving at camp. I was going to walk right by him to get to the tent and grab my sweat shirt but he stopped me. "We're fucking leaving," he said with emphasis on the fucking part. I just nodded and started deconstructing the tent. Joey let me do most of the packing as he already had all of his stuff ready to go, I assume wanting me to feel how he felt having me watch him look for treasure without giving hardly any effort myself. I normally would have been mad, but the storm left no room for unneeded emotions to get in the way. Not freezing in the cold rain and getting out of the forest where our goals could finally align.

We had estimated before we went into Gallatin National Forest that our outbound trek would be about 10 miles using the map scale on Google Maps. What we didn't account for was that those 10 miles were as the crow flies, not as the hitchhiker walks. We still had to hike all the way down this mountain, walk for a couple of miles through a gradual uphill mountain valley, then over another mountain before we would walk our final miles through the forest to get out. Keep in mind, once again, that town was still another 5–6 miles further after we exited the forest. Of course, we didn't know any of that yet, we just knew that it was time to start making our way back to civilization, and thus commenced the longest, most testing trial I have ever been through.

The rain had turned the steeper parts of the trail into a slip n' slide made of mud. Joey and I both ate it and slid off the trail more times than we cared to count. I don't know what it is about getting hurt when it's raining, but it is so much more painful than getting hurt when it's sunny outside. At one especially slick section of trail, I slid off a little faster than usual and reached out to catch myself on a rotting log. It worked, but the short stubby broken off branches cut into the underside of my wrist. My god the little things hurt the worst. The mud mixed in with my bleeding to create a dark, slushy kind of texture that ran down my hand and dripped off my fingers. I did my best to lick the wound clean and try not to let it get to me. We still had many miles to go before we were out of this mess.

Partially due to our sliding, we reached the bottom of the mountain much faster than we had anticipated. Now the hard part began. The stream at the bottom of the drainage had flooded the trail with water that came up to our shins, and while the rain had helped us slide down the mountain, it had also made going uphill near impossible. The steep drainage wall on the other side of the stream was so slick that our first couple of attempts to climb it wound up with the water and mud taking our feet right out from under us, causing us to fall onto our hands and knees in the mud. To make it even better, it looked like some horses had come through on the trail earlier in the day and pooped everywhere. We were literally walking on a trail made of a slurpy shit that we kept falling into.

By the time we reached the top of the drainage to more level ground, we were both coated in a thick layer of mud and horse crap all down our front sides. We were really hoping that flatter ground would give us some reprieve, but the thing about the flat ground is that the water has nowhere to go. We walked through what was usually an amazing green field with tall grass from the looks of it that had been turned into a marshland. We couldn't take a step forward without the mud clutching onto our shoes like it didn't want us to leave. The muddy trail led us through the marsh and more tributaries for miles before we started seeing some elevation change uphill. This was the final mountain we would have to climb before we would start the descent back into civilization. The good news about the unfortunate situation we found ourselves in was that it was impossible to be mad at each other. When the world is kicking your ass and all you have is the guy next to you, there is a sense of comradery no matter how mad you are at each other.

Joey and I working together at long last was not helping my hunger though. I was hitting levels of hunger I didn't know existed now, worse than I had ever felt, without a doubt. Between my body trying to keep me warm in the rain and the never-ending hiking, I was burning through more calories than I ever had before; the problem was that I didn't have any calories left in my stomach. I felt the shift from just hungry to survival mode kick in. My body was pulling from the few fat reserves I had to fuel itself over that final mountain.

A lot of people might not be able to relate to this, but starvation doesn't work like being hungry does. When you're hungry, you just get progressively hungrier the longer you've been without food in a linear fashion. With starvation, it is not the same thing. Starvation is like a staircase descending into the cellar, there are levels that you feel as your body does everything it can to keep you functioning. Those levels are associated with different evolutionary triggers in my experience that took me back to the time of primal man.

Level 1

As the marshland transitioned to the mountain, I experienced level one of starving. I remember the scene clearly because I knew something I had never felt before just happened. We had just turned uphill from a T shaped intersection in the trail. It might as well have been a mudslide as it was impossible to get any traction walking. The slickness forced Joey and I to use tree branches and tiny saplings on either side of the trail to pull ourselves up further. It made for extremely slow progress and took a massive amount of effort. I made the mistake of looking up to see if the trail was going to flatten out soon and lost my footing. Losing my hold on a tree branch, I fell into the mud once again. Catching myself on all fours didn't stop me from sliding back several feet and slicing my pruney skin open on the gravel mixed in with the mud. It was nothing that hadn't happened a hundred times already on this hike, but this time I felt my stomach drop as I pulled myself back to my feet with the help of a tree branch.

I had to let go of the tree branch and fall back to my hands, my stomach felt like it had started eating itself. Not that it was worse than anything I had ever felt, but it was a different kind of pain than I knew existed. For 15–20 seconds, the feeling continued and assured me I was going to die if it didn't stop soon. I closed my eyes tightly shut and gritted my teeth as I tried to let the awful cramping feeling in my stomach fade. In a gasp for air, I opened my eyes and saw the world in a different light. The first thing I remember thinking is looking at the log Joey was sitting on up the trail and thinking, *There might be maggots in there.* My mind was slowly shifting back into the hunter gatherer mindset, everything was a potential food source now.

I had to convince myself that it would only waste time if I tore apart a rotting log when I could be headed to the first restaurant I could find in the next town. Standing up, I was surprised to feel that, while the pain had not left my stomach, it was very isolated and was not affected by other movements. I took another moment to make my peace with it and breathe before I grabbed hold of the tree branch above me and dragged myself up to where Joey was waiting. I couldn't say a word to him, there was a sense of urgency in my mind that I had never felt before. I <u>needed</u> food now.

Level 2

My stomach continued to knot itself up tighter as we made our way up the mountain in the mud that only seemed to be getting deeper. The rain was starting to ease up as we neared what seemed to be the ridge line of the mountain. Because nature has perfect timing, as soon as we exited the forest and into another mountain meadow, the sun burst through the clouds like a reward for surviving the storm. I didn't even look up to acknowledge our good luck, I had to focus on one foot in front of the other or else risk becoming distracted from the task at hand.

Despite the enormous physical change in my body and the pain associated with it, level 1 had come with minimal mental change, other than a desperateness that kept encouraging me to push forward no matter how bad it hurt. Like level 1, I remember the scene vividly as I progressed into my next level of starving, but unlike level 1, level 2 was more of a state change than a painful change.

We had walked maybe half a mile sloshing through the mud in the sun when level 2 hit me. I looked up to see how close we were to the ridge and realized I could see everything. What I mean by that is that my eyes felt like they had opened up to a level they had never been at before. My peripheral vision seemed just as clear as if I was looking directly at something, I was intently aware of movement all around me as the dew drops from the rainstorm dripped off the blades of grass into the soil. I suddenly felt like Joey's boot steps behind me were so much louder than they used to be and I felt I had a much better sense of how close he was to me using just my hearing. Initially, it was very overwhelming. I recoiled and had to stop to be able to process what was happening to me.

What I was feeling was the full hunter gatherer mindset taking over. I liken it to the feeling what it would feel like to discover you have super power aside from the starving part. I did a quick look around me and noticed so much more detail than I normally would. I remember seeing a pink flower bush out of the left side of my eye several hundred feet away, a crow appeared out of nowhere further off in the distance riding a thermal up into the sky, I suddenly had access to such a larger world. My body was directing all remaining resources into my senses so I could find food as soon as possible.

When you are so present, you remember little of what happens around you because there is no lingering on any one moment in time. It is the closest thing to real time travel, one moment I was up on top of the mountain looking down at that town still so many miles away, the next I was feeling the next stage of starvation come onto me.

Level 3

I am sure there are more levels to the starvation staircase, but I only got to level three on our hike out of the mountains. I also believe that if I was to start a fasting diet (purposely not eating for several days), I would not feel anywhere near the things I felt on that day. Sitting in a house and choosing not to eat is one thing, but when you are putting yourself through the most brutal workout you have ever attempted in the cold and rain when you haven't eaten enough for the past three days, it is a whole different ball game.

As with all levels, I remember the scene in level three vividly. We were walking through thick foliage and grass that brushed either side of us as we walked. The tickling grass felt so nice and relaxing as it brushed my arms and calves. I started to feel like resting. *I should just lay down and see if I can take a quick nap,* I remember thinking. Joey asking me why I slowed down was what snapped me out of it. That line of thinking was extremely dangerous, why on earth would I want to take a nap out there in the woods? I think it was my body's way of trying to conserve the last bit of energy I had left.

Over the next mile, my footfalls grew heavy and I spent a lot of time with my head down, using my ears more than my sight to tell me of what was coming next. Everything in me kept telling me to just sit down and take a minute. I knew it was going to be a lot longer than a minute if I listened to that voice in my head. A new wave of, not hunger, but starvation pains started to radiate through my body. These ones were not isolated to my stomach and made me wince with every step I took. My joints started to feel tender and slight even though I had built such thick muscle around them over the course of the trip.

Finally, the trail led out onto a dirt road. Without giving Joey warning, I unbuckled my bag and let it take me to the ground. I marveled that Joey seemed relatively unaffected by the hunger that he had to be feeling as well (I later found out that he had a couple extra granola bars he had eaten before I got back to camp that morning). Pushing Joey out of my mind, I turned my attention back to my body. I honestly felt like I couldn't get up, I didn't have the energy for it. I felt so unbelievably drained from my mind to my tired muscles. I had spent energy I didn't have and, like a bank loan, I now had to pay back my loan with interest.

The Cowboy That Saved Me

Joey impatiently urged me to get up and keep walking. There is no way that he was feeling as hungry as I was if he still had the energy to nag me. Nonetheless, I dragged myself to my feet, feeling my legs tremble beneath me as I tried to shoulder my bag. Realizing that putting my bag on my back wasn't going to happen, I let my bag drop and used it for support while the world spun around me.

I could feel Joey about to say something again when my still acute ears picked up on a different noise. *Click... Click... Click* *Those are footsteps,* I thought. Looking up, I peered through my still sharp vision to spot a person that had me questioning whether or not he was a mirage. The spurs on his boots were what had alerted me to his presence. I had no shame left that would prevent me from begging from this man. As he approached, my first words to him were, "Sir... Do you just...have...a granola bar?" His voice told me that he was surprised, but like any good person would, he said, "Of course. Uh, follow me."

This is when Joey started to piss me off again when he said something like, "O no, sir, we don't need anything." I wasn't sure how much I had left in the tank, but I had a sneaking suspicion that nothing was going to stand between me and the generosity that this man had offered us.

Without even enough energy to throw my bag back over my shoulders, I dragged it behind me in a zombie like state until I reached the small picnic table the man had set up. He told us to sit and he would be right back with food. I got a small glimpse of the treasure trove when he opened up the back of his trailer and revealed 6 feet tall high stacks of freeze-dried meals and boxed foods. He leaned over to a cooler and grabbed some fresh lunchmeat and Hawaiian bread he had gotten only a few hours ago along with a couple protein bars

I felt like I had just been proposed to by the woman of my dreams as I opened the wrapper of the granola bar. Sinking my teeth in to tear off a tiny bite, my whole body seemed shocked and taken back by the burst of flavor. Doing my best not to throw up, I turned and spit the tiny bite I had taken into my hand. *Okay, gotta go slow,* I realized. Taking my time, I felt my mental state restore itself to a less desperate mode.

I paused halfway through my second mini sandwich and had a realization. "If I am coming back to normal this fast, that means that I really was just at the tip of the iceberg in my starving." At first, guilt coursed through me, I wished I could regurgitate the food I had eaten and go back into my starving with my new mindset that I wasn't going to die and that there are more levels to experience. That guilt was replaced with a quiet acceptance that I had just let my own mind beat me. It was a sad and unfortunate failure, but I refuse to lose any sleep over the fact that Joey and I had just hiked through the most intense situation of our lives and came out alive. It was a first step in rediscovering the lost side of us that has been drowned out by the luxuries of modern society.

Recovery

I don't remember much of what the cowboy said because I was so hungry, but I do remember that he tweaked my perspective on something that is broadly looked down upon. The cowboy was a hunter that had traveled the world looking for the best hunts that he could find. Caribou and Muskoxen hunts had been his favorite so far. I used to not be a fan of hunting for sport, but this man started to change my mind. To do most trophy hunting is enormously expensive and most of the money goes to support local anti-poaching efforts within huge hunting 'farms' that promote pristine habitat for the animals that can come and go as they please. The animal's body is also not wasted, contrary to popular belief. The meat goes to local people that use every part of the animal. While I still don't love the mindset of people that want to kill for fun, I respect and appreciate them for funding wildlife habitat for all of us.

We talked with the cowboy for 30 minutes or so and let him tell us his tales while I downed three mini sandwiches. The smile on his face told me that our chat was plenty payment for him feeding us. Thanking him for his generosity, we started off down the long dirt road to the couple of buildings we could see in the distance. Now that I knew I had fallen into a mental trap on the mountain, my trembling legs no longer freaked me out because I knew that most of it was just my mind playing trick on me. Flexing my legs as hard as I could brought the trembling muscles to a halt at last.

Walking through the farmlands was a pleasant change of pace from the muddy uphill nightmare we had just come out of. It was even kind of nice to have the cows moo at us as we walked by, almost like our own personal cheerleaders encouraging us to keep going. We came upon a couple of tiny farm houses that made me feel like I was right at home again. The residents of a couple houses waved at us making me smile. I waved back with all the enthusiasm my still hungry and tired body had left. It seems it paid off when one of the residents stopped us and offered a ride to the nearest restaurant they called, 'The Blue Moon Cafe.' Thrilled to hear someone say there was a cafe nearby, we happily assisted them unloading a bit of firewood from the back of their truck and talked their ears off the four miles to town.

Shaking their hands happy to have made new friends, we jumped out of the truck and ran inside the cafe. We ordered so much food, but the funny part was that we ordered as if we still had normal sized stomachs. It is the weirdest feeling in the world to eat until you're full, but still feel hungry. While I contemplated that new feeling with pie and chicken fried steak (the most calorie rich items on the menu), Joey reserved us a camping site for the night behind the cafe. Everything seemed perfect again, we had food, a warm place to sleep, people talk to. It was everything I wanted, I felt complete. Joey, on the other hand, was far from happy.

Fundamental Differences

I found myself chatting with the locals for most of our time the first night at The Blue Moon Cafe. I was having so much fun and I was so proud of myself for making the brutal hike out of Gallatin Forest. Because I was loving life so much again, I just assumed that Joey felt the same way. I found that was not the case when we laid down to go to bed and Joey brought up that we were close enough to call someone to come pick us up.

181

I thought he was joking at first, but his stern face told me he was very serious. He followed up saying, "My mom would be happy to take us both home."

Taken back that he was not enjoying this as much as I was, I slowly replied, "If you think you need to go home, you should go for it… I am going to finish the trip." That mind shift from Joey changed my perspective on our partnership. I decided that from now on, this was my trip. I couldn't worry about Joey anymore because he was not going to enjoy anything we did. I had to focus on me and what I wanted to get out of the rest of the trip, Joey would have to adapt to my style or leave.

We stayed at the Blue Moon Cafe for another day to eat as much as we could and nourish our hurting bodies. Before we went to bed the second night, Joey told me that he wanted to finish the trip with me. I couldn't decide if I was excited or upset that he had decided to stick with me. I was really looking forward to having a trip to myself, but part of me still felt that we had to finish the trip together. *Hopefully, he cheers up in the next couple of days,* I thought. *It is going to be a very long rest of the trip if he doesn't.*

It was easy to see that Joey had little desire to continue walking when we started off early the next morning. A constant raincloud hung over his head, pouring out all the pent-up emotion that he had yet to release from our breaking apart that happened in Gallatin. It didn't help that he had terrible shin splints that forced him to take long and pained steps. I reached out with questions every hour or so: a joke, sometimes asking if we needed to take a break, but he never took his eyes off the ground in front of him. Everything I said just bounced off of him like I was throwing darts at a brick wall.

Our first ride of the day stopped up the road from us in the form of a Toyota pickup with an enclosed bed. A young man greeted us as we approached his car and said he had room for one in the back and one in the front seat. Joey leapt at the opportunity to use the enclosed bed in the back of the truck and I'll admit, it disappointed me to not have him at my side. It was like another sign that showed Joey and I were not really the team we started out as anymore. I made my decision that Joey needed to go home when the man opened the bed of his truck into the coolest sleeping quarters ever. There were shelves built along the inside to store the little that he had, cool stickers dotted the walls, and the first mattress we had seen in over a week was to be Joey's throne. I was impressed and complimented the man on his style, but Joey only muttered, "Thank you," and climbed into the truck. Nothing about it was rude, but just looking at him, you could see there was no enthusiasm left in Joey.

Our driver closed the door on my old friend and opened the passenger door for me. He was a very easy-going person and drove us the next 12 or so miles to Ennis, MT. I had forgotten the special kind of bond formed between a hitchhiker and the kind of people that are kind enough to pick someone on the side of the road up. I learned our driver was a 23-year-old from the East Coast. His whole life, he had grown up in love with the wild, so it was only natural that he started to feel drawn towards the northwest. Unfortunately for him, his family did not have the financial capability to support him and his dream to cross the country and see the West, yet here he was. He knew that his family would not support his decision to go from the beginning, so his whole high school career had been dedicated to saving for the moment when he would hop in his beat-up old truck and go see what the world had to hold for him.

I was engrossed in his story and asked him, "How long have you been out here now?"

He grinned and told me, "Just under a year." He had landed a job the second he got over here that was enough to support him chasing more dreams like seeing Glacier National Park, seeing Yellowstone, getting to camp whenever and wherever he pleased now that he had so much public land to choose from. Our 12-mile ride only allowed for so much talk, but I shook his hand and wished him the very best of luck on the rest of his journey. I love people that have the courage to chase their dreams so single mindedly. He was living the way of one of my favorite metaphors, 'If you want to take the island, burn all the ships.' In other words, he left himself no option but to succeed, or fail miserably. Now he had built a brand-new life for himself with every intention of rising through the ranks.

Part of me really wishes that Joey got to hear our conversation, maybe it would have cheered him up and convinced him to enjoy the last bit of time we would have together on this trip. I knew that he would have been uninterested though. When he got out of the back of the truck, Joey quietly shook my new friend's hand and looked at me to see if I was ready to go. Well, I wasn't ready. I went to the town store and spent some more time talking to the locals and stocking up on fruit and pop tarts. Admittedly, part of my lackadaisical and slow mood was to slow Joey down in hopes that he would start to smell the roses. I think it just made him more impatient with me. I realized that he was getting flustered and I decided I would try to use it to my advantage. He had all sorts of pent up anger towards me, I could see it bubbling under the surface. If I could break down the dam holding it back and make him mad, maybe I could figure out what his problems were and convince him to stay, but at the very least, preserve our friendship.

I have never intentionally tried to anger someone as much as I did with Joey over the next couple miles. I left all caution to the wind and constantly prodded him in an attempt to get a reaction. I was very clear about my intentions too, not two miles down the road, we found ourselves in our biggest fight of the trip with me telling Joey, "I want you to be mad! You're holding it in, I know you're pissed we didn't find the treasure. Not only that, but I'm pissed at you because you can't seem to find any joy left in this trip!" I could see how mad he was and I could see how much he didn't want to be here right now. If I could just get him to let it out, we might be able to continue this relationship long after we finished hitchhiking. Joey is no easy dam to break though.

One of Joey's favorite sayings on the entire trip was, 'A man keeps his composure.' I always disagreed with that. Look at men like Elliot Hulse, the man I consider to be the best bioenergetics trainer on the planet. He has to release his pent-up rage all the time and even has videos on YouTube of him doing 'primal scream therapy,' which is exactly what it sounds like. Watch him speak and absorb the energy he puts out; you would think he was one of the most composed and masculine people on the face of the Earth.

I use that as a point to show that Joey's 'bottle it all up and hope it doesn't come out' method was not going to work. Joey was really hard to get into, I attest that to a very strong willpower, but he finally did me the favor of showing me what he was really feeling inside. I remember the argument like it was just yesterday:

Me: Dude, you have a rain cloud over your head, you're clearly miserable. Why are you staying with me?

Joey: *(Quiet and intense)* *Deep breath* I am here…so I can keep you focused on getting home instead of visiting every little town along the way.

Jake: Do you realize how selfish that is? I am having fun! There is nowhere else I would rather be in the world but right here, and you are trying to take me away from that?

Joey: *Long pause* You're right…

I had finally realized what it was that was holding him here. Like I had suspected, Joey had little to no desire to continue this trip with me. He thought I was a moron that had betrayed him by not wanting the treasure. It's true, I had not lied but I had also not told him the truth about how little I cared about his one goal of finding the treasure on the trip. That same thing is what was making him stay though! He had been the one that had done our navigation for almost the entire trip, he had been the one that had planned out the majority of our stops, he had been the one that got me off the bus in Colorado when I didn't realize it was our stop. He really thought that if he left me, there was a chance I wasn't going to make it home without him. I really couldn't even take offense to it, he was right! I was just short of completely not caring about my destination and without Joey, it was going to take <u>a lot</u> longer to get home.

What Joey didn't know though, is the chance to fuck the whole thing up and die was what kept hitchhiking exciting for me. I had made a decision that there was only one thing left to do.

Looking over at my longtime partner and the man that had become my best friend, I did something that I knew would be best for both of us no matter how painful it would be in the moment. I gave him an ultimatum "Joey, I'm going to give you a choice. You can either get happy and finish this trip with me like we started it, or only one of us gets in the next car that stops."

I waited for an answer for what seemed like ages before Joey finally responded in a deep voice, "I'll think about it."

We decided to spend the night on the side of the road, but were promptly kicked off by police. Joey and I actually had a good laugh together again at our luck. It gave me hope that he was going to stay now that he had laughed with me for the first time in days. We found a more secluded spot up the road where we could hide from the passing traffic and spent our very last night together.

The Separation

I woke up on the morning of day 29 with hope! It was going to be the day that everything was going to get way better, or way worse between Joey and I. Hope didn't make a difference for what had already been decided though. It quickly became apparent that Joey and I were not going to be finishing this trip together. As soon as I figured out that we were on part of the Lewis and Clark trail, you might as well have considered Joey and I's relationship over. Every couple miles, there was a rest stop with signs about what the trip looked like at this spot or who they met here, my personal favorite was Beaver Head Rock (look it up) where Sacajawea identified it as being near her tribe's summer resting grounds.

Joey was far from interested in Lewis and Clark's expedition even though we had considered ourselves to be a modern-day Lewis and Clark upon the start of the trip. Sighing, I made my decision and asked Joey if he wanted to jog with me. I knew full well that his shin splints had been killing him for the past hundred miles of walking, he said no as I expected. With a last look at my brother, I left him. I jogged probably three miles ahead before Joey showed up next to me riding in a truck. The kind gentlemen picked me up as well and drove us the last four miles to Dillon, MT, the place where it all ended.

It was actually very exciting because I had been in that particular gas station before, making Dillon the first place I knew and had been to before on the trip. Now, I just had to talk Joey into going home. It turned out there was no need for that though, for when I came back out, Joey was just getting off the phone with his grandpa. "I'm done, bud." I felt initial excitement, *I could have this whole trip to myself now!* Then I took a really good look at Joey for the first time in days and felt a wave of sadness come over me. This was my brother, and we weren't strong enough to make it through this thing together.

I took off my bag and gave him a hug. "I'm gonna miss you, man." He was surprised by my affection given how much I had been fighting with him over the past few days. Accepting our hug, he patted me on the back and relaxed. Pulling myself off of him, we made eye contact for the last time. I was hoping I could get some confirmation that we were still going to be brothers after this, but the unspoken language that we had become so used to just wasn't there anymore. Not saying another word, I picked up my bag and walked towards the road that would take me the last 440 miles home. I dared not look back at the Indiana Jones looking man behind me in fear that he would see the tears in my eyes.

The Final Trek
Jake's Final Days Solo

Going Solo

If you haven't seen the new Han Solo movie, spoilers ahead! Walking down the road away from Joey, I felt very much like our situation had become exactly like Han Solo's. In the movie, Han and a woman he loves named Qi'Ra flee from crime lords in a desperate attempt to get off the planet before they are caught. Han makes it through security, but Qi'Ra is grabbed from behind and doesn't make it through. It was heartbreaking for Han, but he had to keep going unless he wanted to get caught as well. Approaching the pilot recruitment station, he begs to be a pilot, "It's all I've ever wanted!" he tells the recruitment officer.

Finally, the officer caves and tells Han, "Alright, give me your name." Han does just that, but does not have a last name to give. Thinking for a moment, the recruiter labels him Han Solo, a name that would later become famous throughout the galaxy. I was Han Solo now that Joey and I had parted ways. I had just lost a big part of me, but in doing so, I was chasing my dreams to accomplish something bigger than I ever thought I would; hitchhiking across the United States.

Home was an easy 4–5-hour drive away on the freeway that cut north, but I was not interested in getting home just yet. I wanted to go south from Dillon to Goldbug Hot Spring and Southern Idaho where I had spent so much of my childhood. I landed a ride in about an hour from a man that had every right to never pick up a hitchhiker again. While we were together, he told me the story of when his dad picked up a hitchhiker last year. I expected some great story about how they were still best friends or something, instead he told me that the man threatened his dad's life with a knife. I asked him right there, "Then why on earth did you pick me up?"

He smiled and said, "I drove by you twice, I just knew you were a good kid."

Awwwww, I thought. Compliments aren't something you get a lot of while hitchhiking so that one went a long way. 30 miles down the road, he dropped me off at the start of highway 324, or Clark's Canyon Reservoir. Despite my driver's obvious disapproval of my decision, I had requested to be dropped off here because I knew it was a much shorter distance to the hot springs I so desperately wanted to visit. What I hadn't taken into account was that Highway 324 is probably that last dirt road highway in the country. There had probably never been a hitchhiker on the 45.6 miles of that road, but I was dumb enough to be the first. There is no food and next to no cars that drive the road. Of course, I didn't know any of that yet, I was much too focused on celebrating my arrival at the reservoir.

Talk about gorgeous, Clark's Canyon Reservoir is beautiful. Much more important for me was that I had taken our dogs for a walk here only a couple months before Joey and I left to go hitchhiking! I knew exactly where I was, and it was an amazing feeling to be so close, yet so far away from my destination now. Looking up, I could see the sun starting to set, which meant it was time for me to find cover. It had rained last night, I had to be prepared for it to rain again without getting soaked. A couple campgrounds dotted the shoreline of the Reservoir that offered the opportunity for a dry place to sleep. Targeting them, I realized I didn't have to agree upon a route with anyone anymore and I was free to go wherever. Smiling at my newfound freedom, I chose the most difficult route possible through the thick sage brush instead of the road.

Arriving at the campsite, I found exactly what I was looking for. An abandoned clubhouse offered a porch with a roof that would protect me from the rain, I would have a dry place to sleep tonight. I forewent the set-up of my tent so I could instead use it for my pillow on the uneven wooden boards. Not a moment after I had set up a good sleeping spot, the rain started pouring once again. I smiled and curled up in my sleeping bag to enjoy the rain from under the shelter. I couldn't help but think that Joey would have loved to listen to the rain with me sitting there on that porch. I even got a little teary-eyed thinking about it and reminded myself that the split was what was best for both of us.

I spent some time in the rain meditating and thinking about all the things I could now do. *This could turn into an actual vacation now,* I thought. There was confidence in the fact that I had hiked through the Gallatin Mountains. Now that I knew nothing was going to be as hard as that, I felt very comfortable taking days to go hiking and enjoy myself now. I only had about $400 left of my original $1300, but it wasn't going to stop me from spending time in my favorite places across my home state of Idaho.

Birds of the Rainbow

I woke up the next morning around 4:30 a.m., to find that it was still raining. The rain had found its way through a few holes in the old roof and left me moderately damp. I was ready to get up, but the cold temperatures quickly shut that idea down, I wasn't about to go anywhere while it was this cold. Scooting further back away from the wet spots on the porch, I had no choice but to sit and wait for the temperature to rise so I could begin my first whole solo day alone. Around 6 a.m., the rain died down to uncover the most beautiful rainbows. I took the colorful skies to mean that finishing the trip alone was the right move. Packing my stuff, I set out in the cold early morning to see what the day had in store for me.

Walking alone is a far different feeling than walking with your brother. Whether we were fighting or not, Joey had always been there to talk to. Now that I was alone, I had no one to speak my mind to except the little birds that fluttered around me as I walked. The little birds were every color of the rainbow I had just seen earlier and sang me all their own unique songs. Alone on that road with those birds swirling around me was the closest thing I ever had to a spiritual experience, because I swear they were encouraging me to make it home.

So much exposure to the only language I could hear prompted me to see if I could learn to sing like the rainbow birds around me. I thought it would be a sort of thank you for all the joy they were giving me on this desolate road that had yet to show me a single other person had ever been out here. As far as I was concerned, it was just me and the birds. For about eight miles, I did nothing but work on my non-existent whistling skills in an attempt to communicate. By the end of those eight miles, I had created my own song, though I had never been able to whistle before those little birds decided to accompany for a few miles of my long trip. With a bow, I sang them my quick staccato song one last time as they flew off into the distance. They had given me all they could, it was time to dig deep and get tough for the rest of the day.

I still had about 35 miles to cover by my estimates and I discovered that I was once again out of food. I laughed at myself and spoke out loud, "I thought you learned your lesson after starving in the woods!" As entertaining as the situation was, there was no laughing my way out of the fact that I was 35 miles away from food with no hopes of getting a ride any time soon. Not to mention, the paved highway had turned into a dirt road. I didn't even know a highway could be a dirt road until I walked on my first one that day. In what normally would have been a pretty dim situation, I actually found myself excited again. This was a second chance at the mental breakdown I had upon exiting Gallatin just a few days ago. This time, I would be prepared to feel the full force of level 3 and potentially level 4 of starvation. With a grimace, I pulled myself back to my feet. I had to make at least 20 miles today if I wanted a chance at eating tomorrow.

Suddenly, I heard tires on the road behind me. *No way,* turning around, I saw a white jeep-looking car putting along the road. Not wanting to waste this opportunity, I held out my thumb, knowing that their slow speed was going to lead to them picking me up. Huge grin on my face and covered with dirt, I awaited my ride. Slowing down even more, the car came to jogging speed beside me and... just kept going? The white jeep picked up speed again and disappeared off into the distance. I couldn't help it, I lost it in a fit of laughter. Talking to myself I said, "Who just slows down? Hahahaha, should have jumped in front of the damn car! That would have got them to stop." I talked to myself until I didn't have the energy to keep talking, I had covered about 12 miles and hardly taken a break so far. It was time for a rest.

Just as I was about to lay my bag down for a nap, another car going about 20 mph putted around the bend in the road. It was an old red truck that I could tell was probably a decade older than me. Whoever was driving that truck was a rancher without a doubt, some of the most trustworthy hard-working people on the planet. They deserved every bit of my respect whether or not they decided to give me a ride. I lifted my thumb up instinctually once again, this time removing my hat and searching for the hardened gaze of whoever drove the old beat up truck.

The Generosity of a Rancher

The car came to a stop at my side, but I didn't reach for the door handle like I usually would. Through the window, I had already found the man's gaze, I was on his time and I would not do anything unless he gave me the go ahead. Holding his suspicious eye contact for a moment, I caught the slight nod of his chin. It was the rancher's language I had come to know from my multiple trips down to southern Idaho that had familiarized me with their culture. His subtle head nod was an invitation to open the door, so I did just that. The first sight that greeted me was a rifle behind his seat and a belt that I could only assume meant there was a pistol on his hip. None of that made me as cautious as his unwavering gaze that stared me down, he was analyzing everything I did, trying to determine if I was someone that deserved help.

For those of you who have not had the pleasure of meeting a real rancher or farmer, let me summarize their character for you. First off, it is an honor to receive a rancher or farmer's help. They will treat you better than anyone has ever treated you in your life because you know that everything they do for you is real, something that has been left behind in modern times. Second, you need to understand that white men on the farm are just as discriminated against as anyone in the United States, but in a very different way. No, they are not called *nigger* or some of the other derogatory terms I had already heard on the trip, there is not a movement like *#metoo* that has been created to stand up for them, but they are the victims of a world that attacks them day in and day out that tells them they are the problem and that they are not good enough.

They get a bad rep right off the bat because almost all of them are conservative, which has somehow become a bad thing in our country. They don't support racism or suppress women being CEO's, they just want to protect the lifestyle that they work so incredibly hard to have. Now every rancher in the United States is labeled an environmental villain by the government and people alike. "Cows are polluting our water! Your fertilizer is causing algae growth in the lakes!" It is all true, these are all things that happen, but because of that they are blamed for almost every problem in the world when it comes to the environment. Here's the funny part, they were never told anything of what they were doing was hurting the environment. They just woke up one morning with a government official on their front step with fines to pay and massive changes to make to their operation that can cost upwards of millions of dollars.

Now it seems the whole world is against them where a zero on his mom's couch in his basement can sign petitions that make tighter regulations for farmers and ranchers that thought they were doing the world a service. Most of them are even happy to change! They don't live in the city because they love the ability to protect wild lands so long as they can protect their livelihood. These people are just holding onto the last bit of a lifestyle that has been passed down for generations and trying to make a little money while doing it.

In that moment on the side of the road, that man looked at me from the point of view of someone that has been hated for years just for trying to make a living. The only thing that got me into that old beat up truck was that I understood that. While most young college kids, including myself, are taught that people just like the one that was now in front of me are the cause of problems; I understand that there is a lot more to the story from my meetings with them in the past.

"Hello, sir," I said to him without making a move to get in the open door.

He responded slowly, "Where are you headed, son?" I told him I was headed to Leadore, ID, coming from Charleston, SC while carefully watching his body language to signal me when to shut up. I could see him visibly relax a little as we talked. I showed no signs of hate and enunciated all my words clearly as to not come off as a spoiled brat that I knew wouldn't mix well with the rancher. "Alright, hop in. I have to check my water up the road so I can take you a bit."

Smart move on his part. The word of a rancher is a vow, he wasn't committing to anything so he wouldn't have to adhere to his own values. I was extremely grateful to be picked up at all, so as far as I was concerned, the man was already a hero. I introduced myself and thanked him again as the rain started to pour outside. He laughed a bit, and before I knew it, the man was telling me his whole story. There are a couple generalizations that go with ranchers. We assume that they are uneducated, and we tend to think that they are uncultured. Maybe it's just me, but almost every single family of the Montana or Idaho ranchers that I have met have completely defied that generalization.

Among the many stories the man told me was of his multiple trips to the big cities like New York and London, where his daughter tended to bounce back and forth. He didn't particularly enjoy the larger cities, but he had been over quite a bit of the world to see them and take care of his daughter. The man was most certainly more 'cultured' than most people I have met.

As we dug deeper into his story, he ended up telling me almost everything, every tragedy and heartache, every high moment and peak. I felt close to this man and he must have felt the same way. Flipping the tables on my never-ending stream of questions, he dug into my story and did a sort of life audit on me. The topic that we spent the most time on was why I was out here in the first place. I told him like I had told everybody, "I was soft and knew I was capable of much more. I want to see what I'm really made of and hitchhiking across the country seemed like the way to do it."

We talked for quite a while on the toughness I was seeking to build in myself, but the one thing I remember him saying to me was, "I think you're gonna be alright."

I hold onto that as the best compliment I received on the entire trip. I had seen the glimmer in his eyes when he said it that told me he thought I was someone worth investing in. We had reached the spot he said he would drop me off and I prepared for the rainstorm outside. I still had 15 miles to walk over the next mountain and I was fully prepared for it to be as hard as the Gallatin hike, but the rancher next to me just kept driving. I looked over at him a little surprised to see him smiling, "Who wants to drop someone off in this rain anyways?" he said. It was my turn to smile now.

He gave me his favorite granola bar to tide me over for the rest of our drive the long way over the pass, insisting that he had to show me a little more about his life isolated from the outside world. We drove by a couple beautiful cabins that his family owned and he told me the struggles that came with building each one, saying things like, "I couldn't have been older than six when we built this one." We drove along a dirt road that used to be a passenger train track. Apparently, back in the late 1800s, investors thought the mountains around Leadore were going to be one of the biggest mines in the world and had prepared for it by building train tracks. I am thankful the plan fell through and left the mountains around Leadore in pristine condition, preserving a large section of wildlife habitat in the lower 48 states. On our final descent, he showed me the section of the Pacific Northwest Trail that ran through this section of mountains. Turns out that I wasn't the only fool walking for miles on end every day.

When we reached the local store, it was still raining, increasing my thankfulness by tenfold as I would have had to walk through that rain with no food had it not been for that rancher. I always wanted to be able to do something for the people that were so generous to me, but after shaking their hands, I would always realize I had nothing to give. Watching that man drive away, I promised myself that one day I would be the one that was giving, not the one always in need.

Familiar Ground

While I had been to a couple places in Montana that we had visited, southern Idaho was a second home to me. I had bounced back and forth all over Idaho for years searching for the best hot springs and now I was back in Leadore, ID! Sauntering down the street in the rain, I found my way into the local store and saw that nothing had changed since I had last been. Ordering a sandwich, I left to scour the aisles for the prized treat that I loved so much; pop-tarts. Not finding them for the first couple of minutes almost gave me a heart attack, luckily a local employee reached out and guided me to the treasure trove. "Smores pop-tarts? No fucking way." I loaded up on about nine of them and ate three before my sandwich was done.

My diet and rough appearance drew in the locals and soon I was talking with everybody in the store about where I had been and what I had done. I had a personal revelation while talking to the locals when they asked, "So, how long have you been doing this?"

I took a moment to count, "23…27… Joey left yesterday. I've been on the road for 30 days." That was shocking to say out loud for me. It had been a full 30 days since I started this trip in South Carolina and now, somehow, I found myself in a little store in Idaho about 2,200 miles away.

"Wow," I said to nobody in particular. I felt kind of proud of myself, but there was another feeling in me that I couldn't shake. While I had accomplished something that less than 1% of the population can claim, I felt as though I hadn't pushed myself as hard as I could have. Gallatin had been one thing, the starving there was unlike anything else I had experienced and had really pushed me past my self-limiting beliefs. The only other thing that had really stretched my mind to the extent of that hike had been our run in with racism in the south. Now that I was eating my fill like a rich man in Leadore, I felt like there were so many more levels I could take this kind of adventure to. As I munched thoughtfully in that little store, I made a second promise to myself. This was not going to be the longest or the hardest trip I was going to attempt.

I left the store with a full belly and plenty of food to last me through the next two days. I left excited, but about three miles down the road, my eyes got droopy, it had been a long day. Sliding off the road into the ditch, I slugged off my pack and laid down in the wet dirt. I never considered myself a good napper, but that ditch was just too comfortable. Before I knew it, I woke up to a sun starting to dip down in the sky.

I laughed at myself, amused that I just fell asleep in a ditch for some untold amount of time. I'm pretty sure that I hit max hobo level at that point on the trip. I was capable of so much more than what my mind would have allowed before the trip. I didn't bother to pull out my phone to check the time, I only needed to know that I still had time to keep walking.

Stretching and making the weird wake up noises that I know everybody makes when they're alone, I dragged myself back onto the road. "Alright, now don't expect a ride, Jake," I told myself. I'd walked almost 15 miles in total and got a 35-mile ride from a rancher, that was as far as I could expect to get in a day. Not 10 steps later, another car stopped right in front of me.

It took a second to register that what I had just said was not going to happen just happened... "Huh," I said, then hobbled up the road as fast as I could in my post nap state. I must have looked pretty funny because the man was laughing when I arrived at his truck. Nonetheless, he let me in and told me he was headed up the road 40 miles to Salmon, ID. Exactly where I wanted to be!

Our car ride passed in more silence than most as the man was one of the people that wanted to poison and shoot every wolf, cougar, and bear in the world until there was nothing but deer left for him to hunt. You can't say much to that level of uneducated selfishness. I do have to give him credit because he was kind enough to pick me up, but I knew any words out of my mouth would most likely end in argument, so I let the ride pass without sharing my opinions.

Ava

As if on cue, when I arrived in Salmon, I received a rare phone call from my younger brother, Sam. Last time he had called had been right before we met Santa Claus on our long walk out of Chattanooga, TN. I put him on speaker phone so he could tell Joey and I whatever story he had planned and it was a good one. He had kissed a girl he hardly knew in front of a school bus filled with 50 kids. I was so proud of him, he always did fit well in the spotlight, but I really must quote Joey to properly describe my brother. Whenever Joey would say anything about Sam, he would end the sentence shaking his head with, "The cajones on that one," (cajones is a Spanish slang word for testicle). I would laugh my ass off every time Joey would say it because it's true. Sam does damn near everything with cajones.

I was already laughing as I picked up the phone thinking about cajones, but Sam was calling with great news for me today. It turned out the same girl he had kissed in front of that bus full of people lived in Salmon, the town I now found myself in, and was willing to take me in for the night. After we got done laughing and reliving the school bus story, Sam sent me her number and told me her name was Ava.

Still giggling a bit as I called the number on the phone, I was answered with a pleasant hello on the other end of the line. Suddenly, I didn't know what to say because my story sounded so ridiculous. "Um, hi. I'm hitchhiking across the country and my brother said I could stay in your house tonight." I thought about the words that had just come out of my mouth for a second and smacked myself in the head. *Maybe a little more subtlety next time, Jake,* I thought to myself. A few seconds of silence on the line passed and I realized smooth was not part of my vocabulary anymore. "Uh huh," the voice said. "Uh, hi, yeah. Have you heard anything about a hitchhiker?" I asked. *Silence* "O, I guess I should introduce myself!" Chuckling, I said, "My name's Jake, I'm Sam's brother. Is this Ava?"

That was the ticket, "O! Yes, of course! Where are you right now?" Not 10 minutes later, a car pulled up next to me and started shouting, "Hey! Are you the hitchhiker dude?" I smiled, tonight was going to be a good night.

Ava and her friends brought me home to her beautiful house on the Salmon hillside where I was greeted by a white goose that they keep in their yard. That is a welcome I don't often get, but I took it for an omen that this was going to be a great night.

Ava's family owns a river rafting company on the Salmon River and manages a lodge miles away from any sort of civilization only accessible by plane. It seems oftentimes while her parents man the lodge in the woods, Ava is left to manage the river rafting company in Salmon. It quickly became clear that she is one of the most capable young women I had ever met of only 18 and more than capable of running the house and doing her job at the company while her parents were away. She is also a hell of a partier and with the house to ourselves, Ava and her friends were running wild.

From destroying me in ping pong to taking me on a midnight ice cream run, Ava showed me one of the best nights I'd had on the entire trip. Retiring to the hot tub at the end of the night to rest my sore muscles, we spent some time talking philosophy before everyone went to bed, leaving me as the last one awake looking at the stars. Smiling at my good fortune, I found a love for the form of travel I had chosen. Just earlier today, I found myself with no food in the rain on a dirt road highway over 80 miles away, now I was living in luxury with good friends that I continue to talk with to this day. I crawled out of the tub to the couch with next trip ideas swirling through my mind. *Amazon trek through the jungle? Australian Outback on foot? Canadian backpacking through the boreal forest?* I couldn't decide which one sounded the best as I fell asleep, but through all of the noise, one thought locked itself into my head, *I've gotta do this again.*

Comfort Pains

Sleeping in is so nice when you know you deserve it. Ava and her friends were already gone by the time I woke up, which left me in a gorgeous house overlooking the entirety of Salmon all alone. It felt like one of those days that I should go back to sleep and relax the day away, but the law of inertia kept me from doing that. Just like an object in motion tends to stay in motion is true, a body in motion tends to stay in motion. I took about five minutes to enjoy the view and got to work with an at home workout. I repped out around 70 push-ups and 100 bicycle crunches, now it was time to step it up.

I finished my sets of push-ups and crunches and moved on to the real test of how strong I had become over the course of the trip. Lacing up my worn-down tennis shoes, I ran out of the house and vaulted off the deck to see what kind of punishment my new and improved body could take. Tucking into a recovery roll from I'd learned from the limited parkour training I'd received in Missoula; I was surprised to find that nothing hurt. No impact aches in my knees or bruises on my shoulders, I felt amazing! *Let's see how far we can take this.* Crouching into sprinter's position, I took off down the road with every bit of fire I could muster.

The wind in my face and the pounding in my chest gave me a feel like Forrest Gump and Spirit the wild mustang had become one inside of me. I ran until it felt my heart was going to explode, but instead of hunching over on my knees to catch my breath, my sprint turned into a long lope that carried me the two miles to downtown Salmon. My god, it felt good to be free. It felt like nothing could contain me anymore because I was a <u>survivor</u>. You could take everything away; my bed, my food, my money and leave me with nothing, but I knew I would still survive no matter what now that I had been through so much on this trip.

Not including rides, so far I had walked around 400 miles on foot. Now, running on the downtown strip of Salmon and racing cars, I could feel all those days walking in the Southern heat starting to pay me back. I realized that the only thing stopping me from achieving anything I wanted is the discomfort and pain it takes to grow. I sped my long lope back up to a sprint and chose a blue sedan as my prey. I had to catch that car. Feeling the discomfort rise in my chest again, I pushed harder. The only way to be sure you are growing is when you aren't comfortable! I refused to let the desire for comfort rob me of my dreams and the life I wanted. It took three blocks, but the car I was racing eventually beat me out. This time, I hunched over on my knees laughing. I was developing a bit of a masochistic mentality that loved the pain of growing. Catching my breath, I stood up and screamed, "LET'S GOOOOOO!" while pounding my chest like King Kong. I was ready to kill the rest of this trip.

I got a call from my brother when I got back to the house that he wanted to come down to Salmon and go to Goldbug Hot Springs with me. I accepted and before long, Sam showed up on Ava's door step. That was a showcase of how close I was getting to home now. It was only a four-and-a-half-hour drive from Salmon, ID back to my hometown of Sandpoint now.

I was happy to see my brother and gave him a hug, but I've learned that it can be very hard to change around the people that you've grown up with. It's even harder to change around your family because they love you so much that they don't want to see you fail, because of that they hold you back from doing anything that might not work out. What people fail to realize is that failure is a necessary ingredient in shedding your old skin in favor of a new and better life. I was lucky that Sam was undergoing a huge transformation as well, so we spent most the 30-mile drive south and the three-mile hike bouncing ideas off of each other rather than trying to tell each other what to do.

I had originally planned on starting towards my next destination of Missoula today, but I felt I could not leave my brother after he had come so far for me. We stayed at the springs for 45 minutes before we jogged back down the trail and drove back to Salmon. As fun as it was to be with my brother again and listen to him talk about all the things he was setting out to do, things just didn't feel quite right. He was kind enough to buy me lunch at a restaurant when we got back, but it didn't even feel right to be eating the flavorful food that came out of the kitchen. It was like things had become too easy all of a sudden and it was taking the wind out of my sails.

I thought back to the moment when Joey and I had sat outside the grocery store in Chattanooga questioning if we would ever be able to come back to the comfortable lives we lived before the trip. Sitting at that table with my brother eating a sandwich and a cup of soup, I had a realization. I was in silent pain at that table from what I now call comfort pains.

Comfort pains can take form in any facet of our lives. You can be a Spartan Race winner three years in a row and still experience comfort pains from the job you can't leave because you're scared that you won't find another one that will pay you as well. You can feel comfort pains as a millionaire raking in truckloads of money every day with no purpose that pushes you to invest it and change the world. Comfort pains are a tricky bastard too. They lull you into feeling like you just need to take some time and re-evaluate what you're doing because you feel like there is a problem in your life when the whole problem is that you've stopped moving! Whenever you have a problem in the head like I did sitting with my brother at that little restaurant in Salmon, the solution is in the body, not also in the head. In other words, ACTION is the best remedy to cure a clouded mind. You cannot rest, you cannot back off, you drive the pedal through the floor and flip your nitro switch to attack everything in your path.

I didn't understand that lesson at this point in my life though. I thought there was something wrong with me for wanting my brother to leave, for not wanting the good food at the restaurant. It felt like for some reason I wanted everything 'good' in my life to go away and leave me alone. I kept thinking to myself, *There had to be something wrong with a person who wants the good fortune to stop, right?* NO. As I've discovered since then, those comfort pains building up in my stomach were only a sign that there was something amazingly right with me that almost nobody else in the entire world understands. Somewhere deep down, I knew I had stopped short of the growth I could have if I kept going.

I didn't fully understand this prospect yet though. The rest of the day, I rode around in my dad's car with my brother looking for something to do. The most fun thing we did was drive up the mountain behind Ava's house and play catch with rocks for two hours laughing like school girls. Unfortunately, we found ourselves back at Ava's at the end of the night. I hated staying in one spot and the comfort pains in my stomach knew that she had already given so much to me by taking me in for a single night. I felt like I had become a freeloader. I watched halfway through a horror movie with Ava, her boyfriend, and Sam before the sickening feeling in my stomach rose to a crescendo. I had to get out of here, but I couldn't leave without my brother who had come so far to see me. I just had to fall asleep so the day would go faster and I could leave early in the morning.

Unable to fall asleep, the night passed slowly. The second that first ray of sun came through the curtains, I was off the couch and on a run down to the local bakery. I grabbed a couple of cinnamon rolls and wrote a note that thanked Ava for her extreme generosity. I couldn't wait for my brother to leave now because I knew that if I didn't take action and start moving soon, the law of inertia would start to work the other way against me. A stationary body tends to stay stationary. By 10 o'clock, I had left the cinnamon rolls and hugged my brother goodbye. At long last, I was finally alone again.

Make Up Time

I walked down the downtown stretch of Salmon one last time and felt disgusted with myself, there was nothing else left here for me and I had allowed myself to stay for two whole nights. I had made good friends, but now it was time to move on. I didn't even want a ride, I had to make up for the laziness I had felt in Salmon.

My requests were granted, from 10 a.m. to 6 p.m., I walked with no rides. That is not to say I got far though. Now that I was on the road, I was much more at ease. I took my time and stopped multiple times to watch the Osprey fly and talk to the Forest Service station I passed about if it was legal to hunt a deer with a knife (definitely the next adventure I want to do). Not to mention that I found myself asleep in a ditch again around 3 p.m. It was an unbelievably lazy day, but somehow, I still managed to make it 12 miles down the road.

'Bad' Kids

A whole eight hours had passed without anyone stopping, but on mile 12, someone finally did. Rolling down their window on the way by, they shouted, "You coming to Missoula with us?"

"Uh… YES!" I ran to the car and hopped in the back with a bunch of kids surprisingly younger than myself. The oldest one among the group was only 17 years old. I could tell that these were the rebel kids that had been allowed to run wild without any structure in their lives. The youngest one was only 13 and even he was taking hits of the cigarette and bowls of marijuana they were passing around. It was not my ideal ride and I had to decline someone pressing a pipe towards my face more than once.

Ten minutes in, I was just about to ask them to drop me off so I wouldn't have to deal with all of their second hand smoke, but things took a change for the better. I started asking a little more about each of their lives, where they were from what they wanted, how they were going to do it, and realized that they were all similar to me. They all wanted more than their lives currently permitted, they all saw big dreams waiting for them on the other side of high school and college just like me. The difference was that they had been told so many times by their shitty environment that they can't do it, and eventually they had started to believe it. Now, they smoked their substances so they could get the high they needed to still believe that maybe, just maybe, that dream life of theirs is still a possibility.

We rode together for a long time with me telling all sorts of stories about how my trip had gone so far. They were entranced in what I had to offer because it showed them that it was possible to get out of your usual life with next to nothing. If I could do it, they could do it, right? I thought I was starting to make progress with them and show them that they could all have very different lives if they wanted it bad enough, but my hopes were dashed when we stopped in a sketchy neighborhood 30 miles south of Missoula. Supposedly, we were, "Picking up a friend," but I saw right through that when another 14-year-old kid ran out of his parents' house with $40 of what had to be stolen cash. The girl in the front seat handed him a bag, he handed her the $40 and we were off again like nothing had happened.

I sighed, I had never pictured myself being part of a drug deal, but here I was. We drove 10 more miles with much less talk than the previous 100. They dropped me off and begged me to buy them cigarettes, a request I eventually caved to despite my personal opinions on the deadly chemical sticks that people get so addicted to. Their mood brightened 10x when I re-emerged from the gas station with cigarettes in hand. They greedily snatched them away from me and I realized that these kids were casualties of a world that never taught them the power of self-value. Lighting up, they disappeared into the night.

Rambo the Mushroom Man

The kids had dropped me off at a gas station next to a local grocery store that I got into just before closing. I managed to buy two burritos before they ushered me out. Munching on a burrito, I walked through the dark and found an empty field down the road to set up my tent. It was just in time too, the rain started falling as I did my best to set up the rain fly before my gear got soaked again.

I dove inside my semi dry tent and zipped it closed behind me. Now, I just had the pounding rain to listen to as my vision faded in the dark night. I once again marveled at how real each day was. There were no time wasters to take away from the presence of the moment here in Montana. Day 32 was over and I was another 110 miles closer to home.

I awoke day 33 only 25 miles from my beloved college town of Missoula. Even if I didn't get a single ride, I could walk that far if I really dug in. Today would be the day I make it back to my home away from home, no matter how I played my cards. When I unzipped the tent, I found that a racoon had helped itself to the half burrito I had left outside. I couldn't even be upset; racoons were clever little devils that were bound to win at least one night against me. Time to get a move on.

It was too early for most of the stores to be open, so I set out on the road immediately and ended up getting picked up by a guy that called himself 'Rambo.' For those of you who haven't seen the movie, Rambo is the most elite special forces operative ever and kills just about everything. To take your new name as Rambo was a big leap of confidence. You would expect someone with such a name to be a military vet, but his beat-up car and undisciplined attitude told me that he was far from it. He struck me as more of a high school jock that had never moved onto the real world. That was a conversation full of possibilities so I asked, "Why do you call yourself Rambo?"

He flashed his nasty teeth at me, "I'm really good at carrying mushrooms."

"Huh?" That was not an answer I was expecting. "Like, what kind of mushrooms?" I asked. He slowed down to 30 mph on the 60-mph highway and pulled out his phone.

"The best kind of mushrooms, man!" He scrolled through his phone viciously until he finally found what he wanted and tossed it to me. I was expecting him to be high on mushrooms in whatever he was about to show me, luckily that was not the case. He had a picture of him with about 10 laundry baskets strapped to his back full of merril mushrooms. I lost my composure and started laughing.

"You pick mushrooms, huh?" The excitement in his voice was unmistakable as he took me through every technique and every location he had ever been, finishing it all off with trying to hire me to pick mushrooms for him. It was a very good sales pitch to be honest, the guy was pretty convincing.

He insisted that he buy me lunch, "At the best restaurant in Missoula!" I didn't want him to spend too much money on me so I tried to refuse, but once again fell to laughter when he looked at me with his mouth agape in shock and said, "Dude! Tell me you've had the new McDonalds breakfast platter!" This guy was too much, McDonalds was his favorite restaurant in Missoula? The guy was crazy and wouldn't stop insisting that I needed to try it. So, I didn't hurt his feelings, I accepted and let him take me through the drive through. When it was finally our turn in line, he leaned his whole upper body out the driver's side window and shouted into the speaker, "YO! You got my McDonald's platter in there?"

Server: Uh…we can make you one, sir.

Rambo: Good! 'O, wait! I need two! I just picked up a hitchhiker and the poor bastard is starving!

Server: O, uh, right away, sir…

I was cracking up in the passenger seat. This guy was the most forcefully generous person I had met. He pulled himself back into the car and nudged me with his elbow, "Just wait until you try this shit, man." We waited through the long McDonald's line to get our prize while Rambo was practically bouncing in the driver's seat from excitement. After a long wait, the servers handed us a plastic box filled with slimy eggs, two sausages, and two coveted hash browns. "Can you believe this shit is only $5 dollars!?" Rambo yelled. I didn't know what to say, but I did know that I did not want those eggs or sausages anywhere near me. I finally got him to leave by saying I was still full from breakfast in the last town and taking his number with a promise to call him if I wanted to go mushrooming. He waved frantically as he drove out of the tiny parking lot with the sausages hanging out of his mouth. I waved back and waited until he was well out of sight before I took the hash browns out and threw the rest of the breakfast away. The important part was that I MADE IT TO MISSOULAAAAAAA!

The Sex Store

I stuck out like Clifford would have stuck out in a litter of dalmatians walking down Brooks Street. Everybody was in such a hurry to get somewhere until they saw me and slowed way the fuck down to stare at me with their jaws agape. Most people just thought I was a hobo, but the town of Missoula is familiar with what homelessness looks like. Everyone knew I was a different breed, with my confident stride and huge backpack. I was one of the adventurers and the people that would slow down to gawk at me knew it. I fed off of their shock and my huge success of making it back to Missoula to fuel my ego to levels it had never been before. When I couldn't contain my excitement anymore, I took off my backpack and just started yelling as loud as I could, "I made it! Me! I! I DID IT!"

I bounced up and down like a fool, but it wasn't enough to burn off all of my excitement. Throwing my bag back on, I sprinted down the next couple blocks with my arms held in the sky like Usain Bolt after winning the 100-meter dash. I held up my hand to a stranger for a high five and the guy totally left me hanging as I jogged by… It seems some excitement you just can't share. I could have kept running, but my chafed shoulders and hips were getting raw again. I took a quick peek down at my hips and saw that they were bleeding again. My shoulders had not quite made it to that point, but the twinge I got when I rubbed them told me that they were close. Blood had become a pretty common part of the trip, but that didn't mean that I wanted pants to match my polka dotted socks from all the bleeding I had done in the South. It was time to give it a rest for a couple minutes.

I unshouldered my bag and hid it in the bushes, there had to be something fun I could find to do while I waited for my hips to scab over again. All the signs buzzing on the street advertised places like Men's Warehouse, Dairy Queen, and Taco Bell, none of which sounded all that exciting for me. Looking up, I realized there was another sign right above me. I walked down the sidewalk to get a better angle and saw that I had dropped off my bag right next to the local sex shop! Back then, I was normally waaaaayyy too shy to walk into a sex store, but I was so horny.

I had been in the trenches fighting through aching legs and bloody feet for over 2,000 miles, and I don't know what it is about me, but there is nothing in the world that makes me want to fuck more than when my muscles are quivering from pushing them so hard. Blame it on being a virgin, blame it on being in the best shape of my life, blame it on my European heritage where every damn country has some story of pillage and rape. All I know is that when I looked down, my legs were not the only thing quivering in my pants.

I looked back up at the electric sign that proudly boasted of their wide variety of hitachi magic wands. I had not the slightest clue what that was, but I did know that I didn't get my wizard letter for Hogwarts when I was 11. Maybe these guys had some answers for me? I proudly stepped inside the store with the expectation of naked women to be sauntering around like they owned the place, but instead had my little boy mind consumed in the hellfire of imagination that fat old men come up with.

The first thing I was greeted with when I walked through the set of double doors was exactly what the entire sex industry is made of; fat old men. I didn't even look at the first product before a 40-year-old fat man with bad teeth was shaking my hand and asking me what he could help me find. I answered honestly, "Uh… I've never been in a sex store before so I thought I'd check it out. What do you have?" NEVER ask that question in a sex store people.

"Well, what are you into?" he asked. I thought the answer was pretty obvious for a 20-year-old kid. *Naked women, of course,* I thought to myself.

Unfortunately for me, the man didn't give me time to answer and started shoving magazines in my hands before I could blink. These were not any sort of Nat Geo or Times like I was used to though. Nope, these magazines were called pornos and the first one he handed me was a picture of a naked chick with a fully erect penis crab walking towards me. I did the same thing any normal human would do; I gagged like someone had put a hand around my throat and took an involuntary step backwards trying not to puke.

Needless to say, my trip in the sex shop came to a quick close. I have never been turned off so hard that I couldn't get turned on again for a full day. I tried too; every image that came into my mind was just replaced with that crab walking monstrosity from the front of the magazine. Try it for yourself, if you get boners in public settings and are looking for a way to fix it, I guarantee chicks with dicks will put a stop to your problem in an instant.

Julien and Jordan

I was back in Missoula at long last, but the tricky thing was that I didn't have a home here anymore. I had lived in the college dorms for the first two years of school, which was great, but it meant that I didn't have a place to come back to in the summer. I started scrolling through all of the friends I had made from athletes to professors and realized that a lot of them might still be in Missoula for the summer. I made my way onto the university campus where I got free Wi-Fi and reached out to about a dozen people to see if they would take in a hungry hitchhiker for the night.

I thought it was a simple request and that my friends would be proud to help me now that I was in need, but I soon discovered that I needed to change my social circle. The responses came rolling in as soon as soon I sent them out with people saying, 'Aw, wish we could! I have some other friends over right now.'

'Dude! Love what you are doing, bro! I gotta work tonight, so maybe next time?' and a bunch of other excuses I was taken back by. It's not like I was going to die if no one took me in for the night, but they really were just bad excuses. I would have driven the three hours from my hometown of Idaho if any one of the people that had just turned me down was in need. I honestly didn't even care about having an indoor place to sleep that night, I just wanted to see some of who I thought were my good friends and talk.

It was a good lesson for me to learn, it's very hard to make real friends no matter what you do. The day that you blow your horn and need help, you find out that there are not many people that answer the call when you just need a friend. It had been five days since Joey left and 33 days since we started this trip. Though I hated to admit it, I was lonely. I had been the untrusted hitchhiker for over a month and now and had nobody to sympathize with me except for the occasional bird that would sing a song for me as I walked by. Scrolling through my phone looking for anyone that might be kind enough to take me in, I found exactly what I had been looking for. The contact read 'Julien.' I sent him a message asking if I could have a place to stay for the night and was thrilled to hear that he would be more than happy to have me.

I had first met Julien on the freshmen wilderness experience that predated our first day in classes at the University of Montana. Julien was a thin and wiry kid that had elected to come to college thousands of miles away from his home in France to study Geology in Montana. Don't let him being thin and wiry fool you for a minute though, he made it very clear that he was not to be fucked with on the first day we met each other. The first thing I remember him saying was, "Fuck." In fact, Julien said some variation of the word fuck so often that I don't remember having any conversation with him at all, probably because of how intimidating it was to hear a thick French accent put an emphasis on the 'u' in fuck. I knew engaging in any lasting conversation with him would end up in me losing somehow or another. His intimidation factor also made him talking one of the most entertaining things ever though; I took an instant liking to him. We had both signed up for a three-day backpacking trip in the wilderness before we started school. The hope was that it would give a sense of familiarity with the university before we were thrown to the mercy of the vicious professors.

As we drove out to the wilderness, our trip leaders took the time to lay out somewhere around a hundred rules that we were not to break no matter what. I didn't come out into the forest to follow rules though, instead I just left my crew in the dust the first two days in the wilderness so I didn't have to put up with anyone telling me I couldn't drink out of the streams. Having people around, especially people with rules for me to follow, ruined the true wilderness experience for me. Luckily, I was the only one in good hiking shape, so I took full advantage of it and raced far ahead of my crew. On day two in the woods, I was surprised to hear footsteps behind me. *Now, who the hell could that be?* I turned around expecting to see someone riding a horse to catch me at my pace, but instead my eyes found a panting Julien working his ass off to catch me.

Okay, this guy had my respect now. He clearly was not in as good of hiking shape as I was, but he didn't give a fuck. He put his head down and pushed himself through the pain to catch me, that is the heart of a warrior in a time where most people are unbelievably soft. This Julien though, he had a little grit in him that I hadn't seen in anyone since arriving in Montana. Ever since Julien caught me on that hiking trail, I knew that he was one of the few people I could count on in my new environment.

I'll always remember us talking about all the girls that we liked while sitting together in the school cafeteria. It had become a tradition of ours to talk about all of our freshmen crushes until one day Julien used his French accent and said, "Fuck it, I'm done. I don't care if I get a girlfriend, I'm going to focus on school and not give a fuck about anybody else." Well, this was a new development in our friendship for sure. What on earth else were we going to talk about now that girls were off the menu? I contemplated this for a full week before Julien and I met once again at our regular time and table. I had a whole list of topics prepared from politics to whale biology, but the stupid grin on Julien's face told me that there was something much more important to talk about.

He said, "Jake, dude... She's in my bed."

"Wait, what? Who's in your bed?" I asked.

Laughing at himself, Julien practically yelled from excitement, "The girl I like! The one we were talking about last week!"

I was so pissed off and happy at the same time. This bastard swears off of girls for one week and falls in love, meanwhile here I am writing a book three years later still single? It turns out the girl's name was Jordan and she is amazing. I thought Julien was pretty great, but Julien and Jordan together have become the most dependable people I know. Probably my closest friends in college actually. To this day, they are there for me on everything just like they were the day I came through Missoula as a hitchhiker looking for a place to stay.

Now that I knew Julien and Jordan would have me if I could make it to their apartment, I sat on campus for a couple minutes and mentally checked off all the people that I needed to stop putting effort into so I could funnel more into my real friends. Just having someone that will lend you a shoulder to lean on every once in a while is huge.

A couple of miles of walking later, I found myself and the home of my closest friends I had made in a long time where they welcomed me with open arms. You know someone is good for you when the first thing they say is, "O, god, Jake! You need to take a shower right now." These people were hard to beat. For two days, they fed me and let me sleep on their couch. That sort of treatment and connection is something that you just can't get with money or fame, this was a true friendship.

I Passed the First Test

On day 37, I woke on Julien and Jordan's couch, gave them a hug, and walked out to the final stretch of the entire trip. 210 miles stood between me and Sandpoint, ID. I did a 300-mile day in Wyoming a few weeks earlier, which meant that today very well could be the very last day I was going to be on the road. I stopped at one other friend's house on the way out of Missoula, but her roommate that I had never met before opened the door. Had to be an awkward experience for the poor girl, but oh my goodness she had a personality. She looked me over and gave me the 'who da fuck are you' look. It was a rough introduction, but she finally told me her name was Danny and let me fill up my empty water pack in her sink. I walked out the door and almost died when I saw Danny give me the cutest half smile ever when I looked back over my shoulder. Women are the greatest thieves in the world as she had just taken my heart in a split second with nothing more than a smile. I had to pull my eyes away from hers and be on my way, I was too close to home to get distracted by beautiful girls.

Unfortunately, the freeway stood between me and home, which meant that I was going to have to wait a long time to get a ride. I waited about 30 minutes and put my thumb down. I didn't really want to go home that bad anyway, hitchhiking had made me feel like I was finally doing something that was making me grow and I wasn't sure home would be able to replace that feeling.

Giving up on getting a ride, I pulled out my knife to sharpen it while I debated what I was going to do back in the 'real world.' As Fate would have it, the second I pulled out my knife, a lady pulled over to the side of the road and picked me up. Now, I have no idea what makes someone pick up a guy on the side of the road sharpening his knife, but I was not about to say no. I jumped in the passenger seat and rode with that lady for 157 miles to Coeur D'Alene, Idaho.

She dropped me off on the outskirts of the city late at night. I had been to CDA a lot and had expected to know my way around, but at night nothing looked the same. Strangely enough though, it seemed I had developed a set of instincts over the course of the trip that just lead me in the right direction. I don't know how to explain what it is like to be lost while knowing exactly what you are doing, but it was a very comforting feeling to have that night. I could feel the change in the air that told me it was going to rain soon; I had to find some cover.

The lady had dropped me off at a T in the road that had a single streetlight overhead. I could see more lights to the left leading to the city, but the right had massive ponderosa pines dotting the sides of the street and led off into the dark. It was hardly a decision to make, I was sick of the city and needed a dark quiet spot away from the cars to lay down. I just hoped I could find it before it started raining.

The dark and the ponderosas enveloped me as I slinked off into the night. The dark used to make me uneasy not knowing my surroundings, now it has become more of a comfort to me than the light of day. Chalk it up to heightened senses or maybe just a familiarity; I felt powerful in the darkness. I knew that no one I came across was as dialed in to the heartbeat of nature like I was. With that knowledge in my mind, I melded into the shadows like I was born there.

The glint of moonlight told me that there was water somewhere ahead and was confirmed by the boat launch sign I passed a moment later. Boat launch hopefully meant that there was a park nearby with somewhere I could hide and sleep. Then I felt that first rain drop on my right hand, I watched it as it trickled down my knuckle before being absorbed by the cuff of my sleeve. I was too late and the rain started pouring down on me. I briefly considered picking up my pace, but I wanted to experience what it was like to be at the mercy of the elements tonight. Being a mere 50 miles from home could mean that I wouldn't be able to experience this again for a long time, so instead of running, I took off my bag and threw my arms up in the sky to let the rain drench me.

I started shivering as the cold water wet every part of my body. I wanted to feel water against my skin so I unbuttoned my shirt and threw it on the road next to my bag. It was just me and Mother Nature now. She had tested me so much over the course of this trip and now I wanted to show her that no matter what she did, I was too hard to break now. The wind started blowing and added to the chill factor like she was trying to remind me what power she is capable of, but I just spread my arms wider and smiled.

The tests from Mother Nature had been the hardest to pass. Not the man with the gun in Nashville, not the racists in Georgia, not the countless people that swerved at us could hold a candle to the humidity that blasted us in the face when we stepped out of the airport in Charleston, the hurricane that followed us through the South, the ant nest I slept on in Alabama, the soul sucking sun that had burned our lips and calves for the first 14 days, the cold winds that stole sleep from us all through the Tetons and Yellowstone, and the bible level storm that we faced while starving in Gallatin National Forest had all been tests that forced us to ask ourselves the same question over and over, 'Are we worthy to become the men we want to become?'

Now I stood in the gentle wind and rain in CDA, knowing that she was only giving me the slightest taste of what she could do. I wanted more, I wanted to feel the hail and wind that would try to talk me into giving up. I was only 50 miles away from my final destination. I could call anyone and they would be more than willing to come get me, but I knew I wasn't going to do that no matter what Mother Nature threw at me. I looked to the sky and said out loud, "I dare you to try." I had challenged Mother Nature once before in Gallatin and she had broken me. Now, it was my turn to take my pride back, this time I knew I could not be broken by her.

As if she knew this, the rain and wind suddenly let up and came to a stop. I was shocked, this was the first time she had ever shown any mercy on the trip. Normally when it started raining, it was an omen for more pain and an even bigger storm soon to come. This was different. There was a sense of completion about the rain stopping that night, almost like I had finally passed a test that I had been taking for the past 37 days. I lowered my arms to my sides slowly and picked up my bag. *It's really over,* I thought to myself, *I passed her test.*

As I settled into bed inside a large bush, I thought long and hard about everything I had just been through. The trip was essentially over at that point; I had passed through the ring of fire and now…where did that leave me? I felt somewhat purposeless now and because of that, I couldn't sleep. I kept asking myself, 'What have I gained from this?' The more I asked, the more I kept coming back to the same answer of, 'I haven't learned enough yet.' The only thing that finally let me sleep that night was when I realized that this hitchhiking trip had only been level 1, the first and easiest test that Mother Nature could give me. I turned on my back and looked up at the sky. I felt like I could almost see her biding her time, waiting for me to take on her next test. Looking up at her, I asked myself, *Am I worthy?* My subconscious rocketed back a response, *Only one way to find out, Jake.* Power bubbled up inside my belly, forcing a smile out of as a single and final thought echoed through my mind:

"This Is Not the End."

Epilogue

I woke the next morning to a chihuahua barking at me through the bushes. I hushed the dog saying, "Hey, it's okay, pal." I didn't think ahead that the dog's owner might not know that I'm here when I said that though. I heard a gasp and saw the dog yelp as the leash snapped backwards, taking the tiny dog with it. I made my way out of the bush to see a middle-aged woman with a dog in her arms running the opposite direction from me. I laughed, "She'll have quite the story to tell later." I put my arms over my head and stretched out a little. I was just getting ready to start my own yoga session when I heard an uncomfortable *cough* behind me.

I turned around and saw a whole group of eyes staring at me. I did some quick deduction and concluded that the group of kids looking at me was a high school cross country team getting ready for their morning workout. Awkward eye contact only lasted a second before their coach uncomfortably shuffled his feet and directed the kids to join him at a different area of the park. It's not often you see a ripped shirtless hitchhiker step out of the bushes at 6 a.m. I blamed their discomfort on my killer 6-pack and giggled to myself as I packed up my stuff.

By 7 a.m., I was packed and ready to go, that left me about 14 hours to cover 50 miles. It would just take one ride to get me home. I smiled, finally accepting that it was time. 38 days on the road had been enough and now I was ready to see my family again.

I set off through the downtown part of the city, grinning as I passed all the places I had been as a little kid. When I looked up at the CDA resort, I saw the hotel floor Mike Fitzpatrick had rented out on Easter so he could give us kids a place to hunt for eggs inside. I couldn't help but grin like a fool; I had so many good memories here. After that, I walked out on the dock where I had dunked my head in the water in December during our high school basketball tournament and sat down. It just felt so weird to be back, but I still had to stay focused. I still had a long ways to go if I was going to make it home today. I bent down and dunked my head in the water for old times' sake. Now, it was time to go home.

With new energy in my veins, I took off at a jogger's pace for the next four miles with my 40 lbs. bag still on my back. I could feel my hips starting to get raw and bleed again but I didn't care, this was too much fun to stop now. I slowed down when I reached the main road that would take me home. It was time to settle in for a long stretch of walking, there is no way anyone was going to pick me up on a road this busy. That's when I saw him.

A grey shirt, black hair, and logger's boots. He was walking down the biking path a dozen or so feet from the road, looking down at his feet except for the occasional glace in my direction. There was a sense of familiarity about him, did I know him? I stopped dead in my tracks and stared intently at the figure walking towards me, his walk, his hands in his pockets, everything seemed like I knew him but he was still looking down. I just needed one good look at his face... "O, my god, SETH, IS THAT YOU!?" I screamed.

The figure looked up and raised his hand to wave, "What's up, Jake?" There is no way this was happening.

Seth and I had been on the swim team together my senior year in high school and become friends in the process. It had been over a year since I last saw him, but there was no mistaking that face. Sprinting to him, I threw off my bag and wrapped him in a bear hug. Seth chuckled, "I knew you were coming home soon so when I drove by, I thought I'd check and see if it was you!" Both of us started laughing our asses off, what are the chances this guy drives by me on the exact day I come home! Catching our breath and wiping some tears out of our eyes, Seth said, "So, I'm guessing you want a ride home?"

I nodded my head, "Yes, please!"

Seth drove me about 35 miles to my grandpa's house where I had stored my car before this whole trip started. I gave Seth a hug and waved goodbye to him as he drove away down the bumpy dirt road. Now, it was just me.

I rang the doorbell and found that no one was home, so I made my way around to the back of my house. There, hiding under a cloth tarp, was my 2002 VW Passat waiting for me to come home. I flung off the tarp to reveal the shiny silver coat of the car that had taken me on so many adventures before this one. I pulled the latch and opened the door to see my keys sitting in my middle dash, begging for me to start up the car again. I took a deep breath, "Time to go home."

I hadn't driven in over a month, so getting the hang of my stick shift was tricky at first, but I had the hang of it by the time I reached the long bridge and pulled off the side of the road. Walking back across the long bridge was the moment I had dreamed of since the day we got driven to the airport in Spokane 38 days ago. I got out of my car and walked to the threshold of where the black pavement turned into the white concrete that marked the start of the walking path on the long bridge. I was one step away from accomplishing the entire trip, one more step and I will have achieved my ultimate goal of hitchhiking across the United States.

I stood there for a long time just staring one step in front of me. Was I really ready to go back to my old life? It was a big question that I just couldn't seem to answer, then I remembered how Mother Nature had stopped raining on me last night and I looked up at the sky again. I knew she was there and it assured me that this step was not the end of our time together. She was just biding her time until I was ready to come challenge her on her home turf again. I suddenly felt sure in the fact that the next time I try something this big, Mother Nature will be waiting and ready to put me through a test harder than anything I went through while hitchhiking. I smiled and looked down at the space between me and completion of everything I had just been through. "It's time to finish what you started, Jake." With that final thought, I took the last step of a 3,016-mile journey from Charleston, South Carolina to Sandpoint, Idaho.

Stats and Lessons

Finished in 38 days.

Walked around 450–500 miles on foot.

Lost around 19 lbs. in 38 days. Started the trip at 158lbs and finished at 139lbs.

Mental toughness increased from around 30% to maybe 45–50% of its full capacity.

Infinite friends made.

The vast majority of people are good.

There were only a handful of people that we encountered that met the definition of "bad."

My mind broke two times. The soccer kid in Atlanta and then starving in Gallatin.

Saved by Joey four times.

7 places that I would like to revisit, four that I know I will.

Too many barriers broken to count.

Love for life increased 200%.

Thank you for reading this story and coming on this adventure with me! It's the first of many journeys that Joey and I will take, so I hope this inspires you to get outside and EXPLORE. Experiences like this are what keep fire in our hearts and make the world exciting.

But you're not done!

Flip the page and start reading Joey's Journal. My partner Joey documented our journey day by day and has a few special bonus stories he's going to share with you.

Enjoy, and prepare for our next book coming your way soon.

Joey's Journal
Daily Journal from the Trip, Complete with Pictures

Disclosure:

This journal is an honest conveyance of my words, and as such includes profanity that some might not be comfortable with. Also, if my words are hard to understand, picture them as a train of thought, or at least translate them into a thought that you might have.

South Vs. North

People Who Honked: iiii – i
People Who Swerved: iiii – i
Who Flipped Us Off (That We Saw): iii – i
Beer Cans: infinite
Cigarette Butts: infinite
Ticks: iiiii – ii
People Who Honked Approvingly: iii – iiiiii
Vuse Cartridges: iiiiiiii – iiii
Blisters: infinite
People Who Had Us Pray With Them: iii – i
Times Laundered: iii – ii
Licenses Checked: ii – ii
Beer Drank: i – i

Bolded = After thoughts

Non-bolded = Direct from journal

Trial Run Notes: May 15, 2018

We walked six hours from Sandpoint, Idaho to Hope, Idaho. My feet were more sore than they ever have been. In a way, I like the pain because it helps to remind me I'm alive. **We left my house at 7 a.m. and followed the train tracks for about 16 miles until Jake's house. Our endurance was so low that we laid in Jake's house for hours after making it. Physically exhausted, and our feet carrying blisters that followed Jake and I across the country, we started back towards Sandpoint at around 4, thinking we wouldn't make it back until around 10 p.m., however,** Dave picked us up on our way back, fellow hitchhiker who had travelled all across Europe and Southern Asia. **Turns out, he lived just down the road from me with his wife. He was originally from the south, and recommended we go to Athens, Georgia. When he was eight, his father had taken his family all across Europe for a year, and since then he had spent the majority of his life overseas.**

Day 1: May 20, 2018

Fuck, that flight took forever. My expectations for South Carolina were way off base, people so far are super nice, or at least tired. I feel confident in our abilities, but I can't wait to be scared. I've never been completely scared in my life, so what I'm looking forward to most is feeling something, maybe it'll ground me. All I can say for certain is I look like an Indiana Jones motherfucker and I feel great! **When we got off the plane, we were immediately hit by the humidity, something neither of us had really had to adapt to in our lives. We bought a hotel room by the airport the first night, and when we got there, it was tucked neatly inside the dense jungle foliage, conveniently just next to the waffle house. We also heard at some point later that there was a city ordinance in Charleston that the building can't be built higher than the church steeple, in other words, there is nothing higher than god. It was kinda neat seeing every building (with the exception of the occasional hotel) being only a story or two high, seemed very classic/old fashioned. Loved it.**

Day 2: May 21, 2018

We felt every inch of the 43 miles we traveled today. Stretch woke us up around 4:50, an hour after we had given up the ability to hold our thumbs in the air during one of our numerous breaks. At that point, our bodies had almost caved in on themselves, and the next four miles to our camp for the night was brutal. We started at six a.m., and after a **complimentary** breakfast, and a few push-ups, we were ready. With confidence in our step, the next five hours only got us 10 miles. We filled water at the local Safeway knockoff, met an older woman, and were on our way for another few miles. Donna was the older woman, but young at heart. Her teeth, it seemed, had seen the light of a thousand suns, **being the color thereof,** and her hug was stronger than her smaller figure would lead to believe. We heard the honk and saw the blinker, and for only a second thought that it was for us. **Then we saw that woman from the grocery store smile from the window of her red hybrid and wave towards us.** She was on her way home, **and although she couldn't take us far, made it a mission to give us her life story and give us each one of those death grip hugs before driving off again.** Sweet gal. 20 miles down. Then down the road, we went again, blisters popped, almost two miles passed before the next hitchhiker picked us up. Both Donna and this man had hitchhiked in their formative years, and to our knowledge, only those who had shared the experience could express sympathy towards the plight of two young men on a mid-day summer Carolina road. 35 miles down. He dropped us off at what he said used to be a pretty rad club back in the day. We ate lunch, **climbed to the edge of the rotting roof,** then continued on our way another four miles until we found a sign to sit under. 'No Loitering' was all it said, **it seemed as good a place as any**, and as we fell asleep while the passing traffic almost sent us into a trance-like state of sleep. **Stretch woke us up with the best glass of water either of us had ever known,** and as we walked those final few miles of the day, a lady laid on her horn as she passed us. Angrily. What a bitch. **We also tried helping a wounded kitten that we had seen as we were filling our water bottles from the spicket of a closest church. From what I gathered, it had been hit by a car and crawled back to the shade behind the church using only it's front two legs given that the back two weren't working. I**

crumbled up a granola bar, left it where it could find it, tried giving it water too, but what I've also learned so far is that it's best not to look back because more often than not, you can't help a wounded animal.

Thoughts of the Day:

1. Excessive exposure to light changes the melatonin in the skin, thus the progression of skin color to tan, or black over generations, as an adaptive response to excessive exposure to different wavelengths of light/energy in an effort for the body to absorb as much as possible to maybe help catalyze metabolic processes more effectively.

2. People don't like hitchhikers, understandably so. We've walked through a very racist area with half of the people, if not more, being black, so how can we blame them for not picking up two white boys.

Day 3: May 22, 2018

Two rides and 110 total miles later, we are just as, if not more sore than last night and rightfully so. Jake saw fireflies for the first time, he really got a kick out of that. I made an 'astronaut' Denver omelet **(freeze dried powder mix)** this morning, maybe it was old or something, but god damn it sucked. We walked from seven or so until about 9 a.m., and right as we were hitting the first hurdle of being sore, and really trying our hardest not to give up after constant rejection, Timothy pulled over and lifted our spirits again. Although he only took us six miles, he revived our spirits **not only with his story of hitchhiking from L.A. to Alaska and back in three weeks, 6000 miles total, as well as from coast to coast, but also told us the value of cardboard signs and how useful they could be.** We filled up our water at the gas station next to the interstate where he dropped us off. **He was on his way to go golfing east with his friends while we wanted to continue north.** Everyone was super friendly and we finally started to hear the southern accent. I do feel bad because a man walked next to me as he was going back to his car, and even though he had ample room to walk around, I still wanted to make sure there was enough and took a step out of his way, regardless of his skin color. Jake walked in an armadillo today as well. **Not on, IN.** Half decomposed. Open walled shoes. No socks. I almost fell down laughing. Our will was tested hard today too. There were moments when Jake and I knew we could do it but couldn't bring ourselves to carry on, **forcing ourselves to keep putting one foot in front of the other.** We made signs at the gas station then walked the next 10 miles-ish with at least one break per mile. I keep asking myself why I'm out here, and with all the cliché excuses I give everyone, the main reason is to find myself and find the willpower that I so desperately need at this point in my life. I need a reason to care. Then Bill picked us up around three. Right before we were about to take another break, a truck pulls over just down the road by the next house, and the man inside gets out to rearrange his things. I was completely unaware that he was there for us until Jake pointed him out. We climbed in and he took us the next 50ish miles to Columbia, SC. Bill hitchhiked from California to Woodstock, which coincidentally enough there was another man at the diner where we were dropped off who had done the exact same thing. I felt bad for not inviting the retired New York post-master in to eat with us. His wife was 'out of town,' his dog had just died, and he was going up to his lake house alone for a while. He retired at 55 and left New York to buy 28 acres down here, **and in his words, it was dirt cheap. From the sadness in his tone, in retrospect, I believe his wife might've left him and he was considering taking his own life that week.** It's okay though, because we talked about death, a topic which nearly everyone who has picked us up thus far casually discusses, **giving you the sense that they were very comfortable with death, possibly may have even wanted to die.** Even Stretch from yesterday told us everything about his grave plot for he and his very much still alive wife. Anyway, Bill said he was a fan of euthanasia, but that he would want to throw a party at 80, then do it. So, he's good. We walked the final six miles from the diner to Jake's Mom's cousin Bobby's home. Sweet and incredibly gracious wife, great kids (except the little one Grayson who kept going on about wanting a popsicle, only to drop his first one and require another) but great family nonetheless! Such a gracious host in fact that he even offered us a ride in the morning. **I cannot express my gratitude for that family enough, they were simply a well-rounded, hysterically funny group of people.** Jake and I are considering buying a bus ticket across the middle of the country/desert area. That's about it.

<u>Day 4:</u> May 23, 2018

Bobby drove us 33 miles to the next town this morning, such a fun dude. We stayed up a little late last night swapping stories and laughing; good family.

Morning Thoughts:

Look at a light bulb in a closed room. Now, imagine the room blanketed in a thin layer of smoke. The light given by the bulb starts at the top and expands perfectly and gracefully down to the ground, creating a cone shape of light. This shape either gets wider or narrower based on how dense the smoke is, and is assumed to be 'refracting' against the molecules themselves. If light keeps 'refracting' **in a cone proportionate to the amount of mass it interacts with,** then potentially at some 3-Dimensional point, light would spiral to an energy state of zero, could this be how dark matter is made? **Given that light carries mass as well in the form of neutrons/ electrons/ protons, it would almost be as if the residual 'mass' left over after all 'energy'/light is transferred would be what we refer to as dark energy and dark matter. With mass, there will always be kinetic and potential energy. What would happen if all potential energy, light, is gone yet the matter remains?**

The minimal energy of light required before it 'refracts' to a state of zero would be equal to the minimal energy needed to create light **for any given mass. If you can replicate this minimal energy for a larger mass, with some proportionality with respect to the change in energy needed to create light, this could be the minimum energy required** for light travel.

After he dropped us off, we just kinda sat there for an hour, resting, taking in the day. We did not sleep well last night at all, too hot and too cold all at the same time, what the hell? We walked the next few miles through and out of town, when just about that time, the cops had received a few calls about us and came to investigate. We introduced ourselves and went on our way. Jimmie came by shortly thereafter, bestowing upon us the classic swiss cheese smile we had grown to hope for from the South. After he dropped us off at the end of his driveway on the state highway, we were easily 40 miles into our trip at 10 a.m. We took another very long break, feeling good and confident, but then came the next 6–7 miles of walking and holy hell was it hard to keep goin. We took so many breaks, and after blistering sun and hours of rejection, we finally made sight of the local collision repair. **Coincidentally, it just happened to be at the end of a two mile stretch of road along the state highway, so after seeing it we still had another good half hour/45 minutes of walking, and this would also have been the only stop for the next 20 miles to town.** Edgar was waiting in his shop, and from what I gathered later on, was one of the best mechanics in the state, even having been assigned to working on the state's police cars. He pointed to a sink where we freshened up and filled our nearly empty water supply. **Walking over to it, never had I felt more like a man, who had taken his reality by the reins and was experiencing something deeper. I felt like a cowboy washing his face in the basin, looking into the dirty mirror and hardly recognizing the sun-beaten grizzle staring back.** Another great guy. As we walked a mile or so further down the road, he even drove up to us, bringing snacks and sodas saying he would give more if he had more, and **although we tried to refuse,** we smiled, thanked, and waved as he drove away. Accepting the gift was the best mistake of the trip thus far. As we laid under a tree and contemplated how to become ecoterrorists, apparently we were on the side of the telephone line that was technically 'not state property.' As we were dozing off, here comes the sheriff, lights flashing, asking for our ID's. He's running them, then the lieutenant comes and we acquaint ourselves. Then another, then another. Four men were standing around us, but Jake and I looked at each other and knew we had it handled. We divided and conquered and by the end, everyone was all smiles and shaking hands. **They offered us a lift to county lines**, we agreed happily. I rode with the lieutenant, and Jake with the first sheriff, all the way to our goal for the day, McCormick, 21 miles away. These men were loaded to the brim with combat gear, batons, kevlar, shotguns; they really meant business. **Could only imagine the daily life of a cop in such a peaceful, yet subtly hostile area.** I got to talking with the lieutenant, and he said that although only three murders happen every year in their county, everyone hates cops for not handling situations better. **For some reason, when we talked about ecoterrorism, I started off on a rant with Jake about the internet and technology, and I thought it was important, so I made an aside in my journal.** Ask me about the internet and I could talk your ear off. Or even technology for that matter. **The offer stands for anyone who wants to discuss the benefits of the internet on our present society, does it make us smarter? Or does it dull us with a bombardment of**

unrelated news and the self portrait of shame we paint for ourselves by constantly comparing ourselves to fake lives of grandeur perpetuated by such social platforms meant to bring us together? The first sheriff, as Jake said, was a great guy as well, they were talking about how there's roughly 10 million cops (probs less) in America, but it only takes a handful to really screw it all up. They dropped us off, 60 miles done, two o'clock, we didn't want to stop. Jake got snacks, then we started off down the road again, only to be immediately picked up by Carol, a lovely Mormon woman originally from Ohio. Her husband was creating a music festival in McCormick, and she volunteers at her church, damn near one of the sweetest gals in the world. She drove us around one state park **on the other side of town because it was close to the road we were going to take,** we were hesitant to get out and start again here, so then she offered to take us to the next state park 13 miles further down the road, so thoughtful and gracious. We considered putting up hammocks next to the river, after all, we had gone 70 miles already.

Fuck that. We kept walking. After multiple breaks and revelation-level conversations about the three consciousness levels;

Subconscious – Lessons from the past that we all use to prepare us for the present and future, and creates desire from what we've known.

The Higher Self – The guiding hand of the mind, intuition, that gives you direction to grow as a person for the future.

Conscious – The ability to comprehend and use both others simultaneously.

We found that more often than not, right when we started to lose faith in humanity and thinking someone might stop for us, literally the next car picked us up. William Perry. Before you even got in the car, you could tell the man's character in a heartbeat. Big bushy tobacco-stained beard, long white hair, a strong face set with sad, heavy eyes. He admitted to being one of the first snipers in the north-western area of Vietnam. He said he had killed many times for the country, and that from his experience in the two seconds it takes to pull the trigger and watch the body fall, he knew almost instantaneously that he had caused grief for some mother, wife, or child. He remembers each kill at the start of every new day. He never told anyone for 50 years after, he was not the type of man for glorification or ratification. He was ashamed of some of his fellow soldiers for profiting off of the lives and deaths of others, and for them turning around and selling their stories. For a man with more control of death than most ever will, also too did he have a clearer perception of love. He made sure we remembered three things:

1. Become informed on your politicians.
2. Love your wife.
3. Love your children.

Sorry if these philosophies might not sound politically correct, but this is from the prerogative of an old Southern man. Love each as if they were family. Love all young as if they were your own children. Just love. He did what he did with the mentality of saving the life of the man next to him, with love in heart, not fueled by hatred of the enemy. **One more reason why war must be a last resort. We must refrain from acting on hate and fear, instead act on self-defense with a response brought on by love for one's brothers and sisters, and commitment to their safety.** For he, too, knew that the other person across the field was just like him, fighting a war that neither had a part in starting. He dropped us off at Calhoun Falls State Park, and the name is misleading. The falls hadn't been around for years after the Savannah river was dammed up. Handshakes and a cumbersome goodbye later, we walked back down the state park road to the highway a little, to a seemingly abandoned housing development **(where the roads and sidewalks had been paved for a few blocks to park boat trailers, there were no houses, the paths were overgrown, and even the gate looked like nobody had touched it in a good long time. At one point in the night, a car drove down the roads, and another relatively soon. We didn't see the cars at the far end, but they must've done some sort of drug deal)** called 'Sanctuary' cuz screw paying $8 each to sleep in a state park. Goodnight.

Day 5: May 24, 2018

It's only been four days and we've covered damn near 250 miles. We slept in this morning, which was much needed, but I've been noticing that Jake keeps waking up instantaneously a few times every night. I think he might be a little scared inside, but definitely putting on a brave face. He's also very tunnel-visioned when he gets focused, but it's bearable. On our way out of Calhoun Falls, we were a walking spectacle for the town, at the gas station, and at the diner. Everybody is so damn friendly here! Except when you're trying to hitchhike. We walked to the Georgia borderline, which was iconized by a beautiful bridge over a gorgeous crystal-clear lake. Islands of clay added freckles to the water and a few fishermen made the atmosphere that more inviting. Jumped off the bridge. Naturally. Then only after four more hours of walking did we get our first ride. Damn near every time we seem to completely give up faith in the chance that someone might pick us up, the next car is the one that pulls over to give us a ride. If that isn't symbolic in any way, then I don't know what is. A true test of will. My calves to knees are so damn red and tender it's crazy. Mark gave us a ride, off the state highway and through winding country roads to the next town where we filled up water again and thanked our luck that we got off that god forsaken stretch of road. **He then proceeded to have us pray with him for our sake, but not in a commandeering way, just for the thought of safety, and a thought very much appreciated, though we aren't religious men.** Everyone was so damn nice, but also stupefied that we were there at all! We got a lot of laughs, honks, and a lady even took a picture of us **as we set up under a tree on the street corner outside the dollar tree,** but probably for reward money if we turned out to be criminals. 25 miles. Apparently that town was the granite-stone capital of the world. Neat. We set off again, down the busy highway, more honks and looks but still nothing. My new sign read 'Will sing for ride to next town,' and Jake wrote 'Will dance for ride to next town.' Then we figured that with all the happiness we had already brought to those who picked us up, that more people should be encouraged to hitchhike. It's a brutal experience though, tell you what. Plus, hitchhikers are scary, but why? We thought it was because of the unknown, but what if there was a way to make it known and incorporate it into the technological age somehow. Then we brainstormed a company that would evaluate applicants based on their history, and if verified as a decent person with an interview, they receive a hat and a sticker. The stickers could also be given out for free so long as they were verified. You wear the hat while hitchhiking so people know you're friendly, then put the sticker on your car so people know that the person picking them up is also friendly! Then possibly incorporating Google into a hitchhiking map-layer that allows hitchhikers to input their ratings of roads relative to how easily (or not) they get a ride. This allows people to connect and talk to new people in the modern age **in an archaic fashion. Maybe there would be a registration fee that would be used to fund road clean-ups of most popularly visited roads.** Then it would show hitchhiking quality on those roads such as ample side-of-road space, or litter amount. Just as the beginning part of the hurricane passed over us, Joseph picked us up, but of course we had to sing and dance first. He had shown us music that he himself produced. I'm a snob when it comes to music, like I pick it the fuck apart in my mind, but Jake was grooving, and he loved it, so it was alright. He finished his first year at UGA here in Athens but didn't know if he was going back for the fall semester. He had read a lot about hitchhiking and travel, so he was very sympathetic to our plight, even planned to do it himself one day. Good kid, pretty sure he was gay though, kept looking at me often, and had a shit ton of panties in his car, more than likely his, but who am I to judge what makes a person happy. Finger nail polish, a clutter of CD's and magazines full of a car ride later, he dropped us off in Athens, GA. Our destination for the night. 55 miles today. Marc was a middle-aged veteran who did some thumbing back in the day as well. When he dropped us off, he had us pray for safety with him, we obliged, but again it was his journey and his happiness, who am I to stand in the way? We walked the next few miles **from where Joseph dropped us off** at Bishop Park, and plopped down in front of the security office that had a wall socket, overhang rough, and a spicket. We were too tired to keep going and knew there would be no better, so for tonight, we lie outside on the stone ground surrounded by the sound of crickets on a baseball field and our personal tree grove, **complete with a serenading downpour from the incoming hurricane.**

Day 6: May 25, 2018

Today felt more like a blur than anything. Like it's almost hard for me to remember if what happened today even happened today or not. Jake and I got under each other's skin a bit, he's a deaf motherfucker and I hate repeating myself, and my responding tone makes Jake kinda pissy, but we overcame it a bit this morning walking out of Athens, and by mid-day the serious Jake was gone and the crazy fun one was back. He had all kinds of energy and we were racing each other up the highway and everything. First person to pick us today was David around 10:00, but only eight miles from where we started. He took us to a total of 15 miles, which was nice. The man lived alone, said he and his brother hitchhiked from Colorado to New York, and that they ate cow grain on the way to save money. With a tone of fondness in his voice, it almost seemed as if his brother had died some time ago. We expressed our appreciation, and continued for another few miles until we jumped in the bed of Mike and Carl's truck. We never met them the whole way through, but by the firmness of their handshakes, they seemed to be of upstanding character. We walked for hours down the next road, through storm and drop-dead heat, we persevered. At times, it almost seemed like Jake had given up before we even started, or at least just kept resisting his body's natural ability to take over and continue on without much mental effort. Even now while I'm writing this, he kicks me in his sleep. We finally made it to a gas station, and the store after. Jake was walking ahead of me for only a few minutes. In that time, somehow, he re-found himself and was back to being the cheery guy that I almost hadn't seen since we jumped off the bridge. Then only barely the day before that. We walked for another good seven miles, nearly dusk, then from ahead of us comes Jen, Ricky, and their daughter in a beautiful blue ford f-150, from the 90's most likely. They let us jump in the back just after we had fully given up for the day. They took us to the next town of Loganville, **at the minimum mile mark we wanted to reach for the day.** We put up a tent behind the gas station leading south to my aunt in McDonough, just south of Atlanta.

Cardboard Ideas:

1. Your Ad Here
2. We are going the same way as you
3. What would YOUR god do?
4. Please stop flipping us off
5. Will Hitchhike for Ride
6. No Drugs, Sorry :(
7. Sexy ✔ ✔ ✔
8. Threesome for women only
9. Pick us up or fight us

Day 7: May 26, 2018

The road is littered with the vices of man. I can't begin to tell you how many beer cans, alcohol bottles, and cigarette butts we've walked over. After a good night of sleep, we settled into our groove and began walking for a few hours. We were in decent moods after being honest and having told the other what was on our minds. It was good to walk on the same side of the road again too, we had each taken a side on days past, but we believed that this could be the main reason we weren't getting along well; he couldn't hear me and I don't like repeating myself. Patience is key. We walked until 10 then as luck would have it, Jillian and Ricky (not Jen **as I previously thought from yesterday**) gave us a ride to the next town again! Total life saver, her and Ricky met us again after we pulled into the Les Schwab parking lot, they were both ex-military, he was a medic, which to my surprise he needed no experience to qualify for. **I had always assumed one would need some sort of medical experience/knowledge, but it makes sense that they would teach you everything you needed to know. I now have the goal in mind to be a medic in the Navy after graduating college. My dad was in the Navy for 10 years working as a sonar technician on nuclear submarines, and so I go Navy too.** I spent the rest of the day genuinely romanticizing the idea of being a medic, which I really want to be right now. By that time, it was 11:00 and we stopped in at a Hooters, naturally. Boy were we a spectacle, **again,** but to our surprise, it was a very family friendly environment because only half an hour later, it was packed with the livelihood of Snellville. Those dry-rub wings were so damn good, and they threw in a plate of fries for free, so we were having a fantastic morning. We walked from noon to 2:30, but we were so in the moment that it felt like an eternity. Jake and I had become really frustrated with each other up to then, and it didn't help getting smoked-out by some asshole in a big ol' diesel truck when he drove by. I had truly seen a side of Jake I never knew, the focused, serious side that couldn't get over his idea of the destination to truly embrace and enjoy the now. I was trying to direct us to some super old, dope-ass looking monastery, but Wes ended up giving us a ride. First to his house where we enjoyed a couple beers, then to my aunt's doorstep. Jake got to know him as I rode in the bed trying to calm down, **the**

231

beer definitely helped, but from what I understand, he had become a great dude. He lived in a camper as he fixed up a cheap foreclosure house (he had bought) with his dad. Turns out, the guy used to be a high-level meth dealer/addict back in the day and he was so infamous that he was wanted in 6+ states. Then one day, he got a call saying they were coming for him, so he dyed his hair, changed his name, and went on the run for 14 months. One day, he was hiding in a meth house with two other people on the run and their baby. The cops circled them but the other two left their baby, so he gave himself up to make sure the cops knew about the baby, to make sure it was okay. He was tried in every state court he was wanted in, but only sentenced to two years in prison since he was only convicted of conspiring to sell drugs. So, naturally, he read the bible multiple times and really turned his whole life around. We got to Aunt Jenny's around four, hugged, took pictures, all the good stuff, but then she went out to get some food for us. On the way, she was in a car accident with my younger cousin Katey in the passenger seat. With my older cousin, Sean, at a Kenny Chesney concert with his girlfriend and my Uncle Mike going to the hospital to be with them, Jake and I were alone for a good few hours in their home. It was good to relax, and after some pizza, we got an amazing night of sleep. **It's funny how even when there's emotional distress, our bodies and our subconscious truly do control more than we give it credit for. Never had I truly appreciated sleep until I was so physically exhausted that I couldn't help but give in to sleep. I've joked around a bit by saying that we aren't actually in control of our lives, but rather at the mercy of our subconscious. The more honest and in tune we are with what we desire, i.e. our subconscious, then the more in control we are of our lives.** My 12-year-old cousin Katey was diagnosed with ADHD and anxiety a couple years ago, so now she takes pills to help her with it. Bless their hearts, but my family has truly bought into the whole capitalistic/heavily-conservative lie that the south both preaches and perpetuates. Don't get me wrong, liberals are just as bad and if it were up to me, we would do away with the party system and labels altogether. They only divide us as a united people. I don't approve of her taking pills at such a young age. 35 miles.

Day 8: May 27, 2018

We got our fair share of the spread of diversity and the opinions thereof today. After my uncle and Sean spent the majority of the afternoon scarring us with horrible stories of people being murdered and shot commonly and with no cause, **literally just for the heck of killing someone,** and after spending the morning with my aunt and Katey in 'Gritz,' where we all had some good ol' country cookin and real grits for the first time in my life (not bad actually, it's pretty good with sugar and butter) then walking through confederate memorial square, we got a ride through Atlanta from Sean. Aunt Jenny was damn near the spitting image of my grandma, except with an increase of humor, sass, and sarcastic abilities, but nonetheless a very religious, traditional, southern woman with a sense of 'keeping up with the Joneses' in the family. **Ironically, 'Jones' is the last name of those grandparents too, who set the standard for 'Keeping up with the Joneses.'** Nicest damn lady you could meet in the South, **when you're in her good graces that is.** Uncle Mike was a stern man, meticulous in his thinking and instructive damn near to a fault. He could tell you how to work an oven five times in a conversation, but at the end of the day his family came before all else. He loved the hell out of each of them, and I'll bet the large dip of chewing tobacco they both had at all times that he would lay down his life for his family. That's what I respected most about the man. My cousin, Katey, was a sweetheart, and a beautiful little singer. She put a pillow on my side the night before to watch TV, and the whole next day we were inseparable, staring contests, singing together, walking the dog, hitting each other; I really enjoyed her company. It was hard saying bye to her because it almost felt as if she didn't want to stay in the environment she was in because her fun-loving spirit wasn't always appreciated or understood in the same sense that only another crazy person could (me). Sean dropped us off 40 miles from the house around four. We started walking north, satisfied with the idea that we were as far south as we were ever going to be on this trip. **It was a nice thought.** We walked for a few hours, but with the extra car-ride full of scary stories of humanity we just had with my cousin, our hearts were only but a little broken and our perceptions tainted with the possibility that people weren't good at heart. **Conversations are like currency, same with** experiences. The more conversational repertoire a person has, the more their 'worth' goes up for conversations because then they can talk to more people and network and blah blah blah. The conversational currency of the South is fear. People perpetuate the idea of fear by spreading their knowledge of scary stories and biased perspectives on reality. We were so bombarded by how many stories we heard in such a small amount of time that we were literally struck to our core on how hateful a region could be. Even if only a small amount of that region was biased, the bias spreads fear and that fear is what prevents us all from becoming the people we want to be. I shit you not, there was a legit confederacy memorial, complete with flags, statues and flowers. In all fairness, it was a beautiful plaza, whoever constructed this pro-racist history memorial did a spectacular job. The flag is indifferent; it's a piece of fabric, but the eyes of the person who waves it speaks volumes. I don't care if the people feel like 'the confederacy is a part of their history' the biggest motive of the movement was literally keeping slaves. So, excuse my language, but I am so fucking tired of people making excuses to hate. Did any other country keep memorabilia from hate-filled chapters in their history? No. No other country except for those currently war-stricken retain the dark chapters of their history. They learn, then move on, like how people were meant to live. Look at Germany, it's illegal to produce, distribute, or display symbols of the Nazi era — swastikas, the Hitler salute, along with many other symbols that neo-Nazis have developed as proxies to get around the initial law, and if charged with any of the aforementioned, one would be looking at five years in prison. I'm not saying we need to go to that extreme form of punishment, freedom of speech is fundamental to the United States, but we need to take a hard and honest look at ourselves and how we conduct ourselves as a country. Let's move on as a country. We were left in a predominantly white town (my cousin did this on purpose because he knows the ethnic-regioselective intricacies. It sickens me that this is the truth), and as we were walking, Jake's 'Please help us get home' sign got some unwanted attention from an older white woman who wanted to give us what looked like $100. We knew damn well that if we were any other color, she wouldn't have done the same, but the worst part is the look she had on her face; the 'holier than thou' smug face. That is what broke us. We both damn near felt like crying then and there with all of the hate and hypocrisy surrounding us. In a bible belt region, it's disheartening to know that so much bias exists towards those of different colors, from white and black alike. It's not a one-sided hate, and for supported reasoning from crime from both, but like Jesus, turn the fucking cheek. But we kept walking, talking, analyzing, and trying to find the good in the world. We talked with a young kid waiting for his mom soon thereafter, he made certain to vocalize how crazy and unsafe he thought we were. We took a minute, then came back at him with the idea, 'What kind of life is a safe life?' **In all honesty, we're made to want a safe life, and everyone who loves us wishes that for us as well, but isn't safe just a nice way of saying you're not willing to take risks? Life is all one big risk, there is no true safety, only the illusion of safety. To get ahead of this illusion, we must take the reins of our life and roll the dice every now and again. I'm not saying hitchhike around the country, nor do I recommend it, but safety is the inhibitor we must unmask first to claim life as our own.** One family rolled down their windows and gave us props for being 'real adventurers,' saying they were planning on going to Colorado too, but that mostly they were jealous of us. That lifted our spirits. **When overwhelmed with a lack of empathy, it makes one appreciate even the smallest support even more. I live my life looking for a balance in all regards. You cannot truly appreciate having something unless you can appreciate not having it as well. The yin and the yang, the dark and the light, the hate and the love. I love love, and I hate hate, in all extremes. There is a duality in every mentality and issue, and finding the duality in things that interest you or pertain to you are very worthwhile to truly find appreciation for it.** We made new signs. Both read 'Traveling Country,' but mine says 'with love' and Jake's says 'via generosity.' We got a ride just up the road for about 10 miles from Zachary, who had just left his three children, dog, and newly divorced wife to go back to his now empty home. He took us well past his road, and in return, we gave him an ear to listen to. **Jake and I had worked out a system by now to switch who would sit up front every other ride to entertain our host without one or the other getting too worn out by having to talk to complete strangers while the other doesn't.** It sounded as if he took his anger out on his wife, possibly even hit her a time or two if the situation was right because he said he thought of himself as a relatively scary/bad person. He showed us his knives, asking us if we wanted one of them, almost as if he was considering killing himself and wanted to give his things away, but this was just a man who did not fear death. Experiences do not define a person, character does. **Character is how we learn from our experiences, and it takes an objective perspective to learn all you can from an experience. Actions speak louder than words. It's how you carry yourself in the now and every second thereon that matters, so long as you keep trying to learn and become a better person from every experience.** We walked the next few miles to a gazebo up the road and on the property of a very capitalistic looking church, we can only assume Christian.

Also adding to the hypocrisy of the bible belt, practically a mansion that 'praises god' and yet I've never come across a deity that approves of materialistic motives. There's literally a deadly sin named after money; greed. It's cool walking through a lot of these old towns because this whole area is stuck in the past, the buildings are decades old, and it even seems as if some of the factories are from when the country started. But everything is stuck in the past, even the majority of the people.

<u>Day 9:</u> May 28, 2018

We made it out of Georgia today, and honestly I couldn't be more grateful to get out of there. Easily the longest five days I've spent in a long time. **Time in itself is a currency too, it reflects our interests in how much time we pay for our interests, which in turn reveals the characteristics of a person based on how they spend their time.** The people were amazing and sweet, don't get me wrong, and the food was heart-stopping to die for. **So much fat and sugar, it was great…** But damn, are the people proud. Jake and I not only found that the entirety of society (or at least damn near) is so afraid to step out of their comfort zone, which rightfully so as a result of our leaders and 'new' sources bombarding us into being scared. Crime and hate are sexier news stories and attention grabbers. It's not like crime didn't exist before TV/internet news, but now it's 'suddenly' increasing and we're made constantly aware of all the despair in the world. But also, the people are so damn proud of their heritage, town, color, blood. **The man who raised me is my father, and yet we are not related by blood but by a bond much deeper. He has taught me tolerance, love, strength, and has given me a strong role model to base my life around. Blood is superficial, and family is what you make of it. By embracing this mentality as well, it makes love for one's friends, brothers, and sisters more attainable because loving your family is rewarding in so many ways.** They hold so fast to their history and ways of the past that they never truly open themselves up to new ways of thinking. Ways that can progress our species to the forefront of not only our survival, but mankind's manifest destiny of the universe. We woke up at 6:30 a.m. to the drones of cars flooding into the church on Memorial Day, as if they weren't already godly enough. They were probably decent people though, I've just been in a critic-based mood lately. We left at seven because we knew we shouldn't taint the gazebo of god any longer than we needed to, and walked to the gas station just down the road to prepare for the day. Lucky enough, while we were there in the morning rain, a rookie officer was on his way home from the gym (which was an hour away for some reason at the asscrack of dawn, so whatever). Long story short, I don't believe he was at the gym. Isaiah, 23, took us the next 35 miles on his way to Rome, GA, which was right along our route

238

since we were trying to make it to Chattanooga, TN for the night to see Jake's aunt, an amazing woman and amazing chef, by the way, who teaches cooking classes but only as an intense hobby. He told us all about his trip to Malawi for seven months of mission work and all about his landscaping business venture, finally settling as a cop for a bit until he finishes fixing a textile mill and converting it to a rock-climbing gym. **What a neatly buttoned life.** He's been married for three years as well, with a little one on the way. He was a clean-cut, eagle scout of a man, very soft-spoken, but firm. His presence demanded respect, although he kept his car a bit messy. Nice enough guy to drive us all the way, and even prayed with us **in the IHOP parking lot.** We let people do what they do for their own sake, live and let live. **Jake and I have never been religious guys, but we do see the value that religion plays in helping people cope with life. The promise of something after death is too exciting not to believe in, and ideology in general helps the entire world every day, giving people purpose and providing them with upstanding morals to hold themselves with. However, it does start a lot of wars. Since the beginning of time. So, there's that but whatever.** Wasn't the first time, and probably won't be the last. We're still out-running the head of the hurricane that hit the South-East coast, **that was just the beginning the last few days leading up.** So, we started walking north to a fancier fast-food establishment called Jack's. All of the ladies there couldn't get enough of us, and I was doing my best to flirt it up and get their day going right. Great food, and another great thing about the South is that since they're still experiencing an economic slump, everything is so damn cheap! Since we already hit our daily mark at 10:00 a.m., we decided to wait out the rough part of the storm and we called a gal named Caity, who is considering joining us on the road sometime later in the trip, fun gal, and just from talking with her on speakerphone, we could both tell she would make a fun addition. When the rain slowed again, we zipped up our coats and kept walking until three. Jake took us down a great back country road, where we saw bamboo, oddly enough, turns out they're an invasive species. Obviously, as grown ass men, we sword fought, then went and fed some horses. We were scouting a good sit-down spot up the road when (and I'm finding it harder to care enough to remember some names, but for sake of recollection let's call him Rupert, but I remember his name sounding just a little dorkier) Rupert and his best-friend picked us up in their new ford that they were taking for a joy-ride. Beautiful Black Beauty, almost felt bad about getting inside it was so new. Jake talked their ears off about local wildlife, the Appalachian Trail, and about some of Rupert's best-friend's stories where he almost killed a deer with a bowie knife (Jake's new dream. **Not for the killing aspect, but for the ability to best nature in strength by hand**). First, they were only taking us to the county line, but since the next town was so close, they said fuck it and kept driving there, and then the interstate was just down the road, so it turned into another 20-mile ride. Both were big men, also both were on the county sheriff narcotics unit. Very typical southern country boys, chew and a spitter, but nice as hell and they got a kick out of Jake. Turns out, Jake was more conservative politically than I thought, mostly on tax issues, which is okay, but still, learn something new every day. His Aunt Dawn, being the generous woman she was, insisted on getting us from that exact spot only an hour later and drove us to Chattanooga, TN, another 40 miles away. Luckily for us, in the time we were waiting it was raining harder than it ever had, and for Rupert, we were grateful. **Have you ever watched how people act in the rain? Why do they scurry about so goofy trying to get out of the rain and keep their heads down? It's not like you can get any more wet than you already are, so why even try to fight something that's going to happen? If anything, people limit themselves by not being open to appreciating the sheer power and beauty of the weather. If you don't keep your head up and aware then you will never be able to know your surroundings and could fall victim to something you'll never even see coming.** We stuffed into her mini-cooper and chatted it up. She seems a little lonely, and really loves the company of others who were willing to listen and reciprocate decent conversation. She was taken aback the rest of the night as we cleaned up after dinner and made sure she was well attended to, saying she never gets this kind of treatment. Sad for a woman so giving and self-less. Fuck, can she cook though. She made banana flambé over vanilla ice cream for dessert, and although it only took 10 minutes to make, it was so damn good! She knows how to cook it all! She gave us the loft to sleep in, set with a king size bed, Xbox, and goodies, saying we can stay as long as we'd like, but we almost feel uncomfortable being here because the whole point of the trip was to distance ourselves from such luxuries. Nonetheless, we are enjoying it to the fullest, **just as we try to do every day of our lives.** It almost feels like the Land of the Lotus Eaters from 'The Odyssey' in a way, because it seems as though all of these luxuries are designed to make us want to stay, but we have to convince ourselves that this isn't in our best interest. My blisters smell like death and I really don't like how my feet smell like that of my biological father's. It's unpleasant to say the least.

Day 10: May 29, 2018

I'm writing this on day 12 doing some catch up, so I can't attest to the complete accuracy of my words. We slept in pretty darn late that morning. The Land of the Lotus Eaters definitely got the best of us. We woke up to candied bacon and quiche in bed. Brown sugar and cyan, 350°F until crispy. From there, we had an amazingly lazy day, to the point where living in luxury was uncomfortable and we couldn't wait to get back on the road. Lunch w/Dawn's youngest, then Jake and I made downtown Chattanooga our bitch for the next five hours, climbing on as many things as we could, and Jake was doing flips off of everything in sight. That's one of the things I appreciate most about my best friend, his capability to look at something and see it for more than it is, but rather for what it could be. **We walked through a lot of businessy historical looking buildings just for shits and giggles, and I remember in one of them there was a wishing fountain. When I was taking a penny out to wish it back in, an older, larger, cliché business woman (probably named Ethel) grunted an 'NUH-UH' from the door of her business down the hall. She didn't bother to get to know who we were, or why we were there, she only assumed the worst and displayed disapproval. She didn't even know that all I was going to do was throw a penny back in and wish for whoever threw it first to get their wish.** Phoebe was the one behind the ice cream counter in downtown Chattanooga, she was a sweetheart. Then we walked down a neat, rustic-artistically made, blue bridge that stretched across the river there. **I was ecstatic about walking around Chattanooga, there was so much history all over town, as well as a beautiful art museum that clung to the top of a cliff next to the river. I somehow lost Jake and Dawn just outside of it and waited outside for about half an hour. Little did I know, they were inside the museum looking for me, and they were able to convince the security to let them in for free to look for me. Motherfuckers took my museum experience XD (that's a sideways laughing face for those who don't know how to read text-talk).** We took a tour of Dawn's Christian radio station, which happened to be the most successful for the region. Then we hit the store real quick, and proceeded to go back to Dawn's to make a 'craft your own spring roll bar' for dinner. **Another great thing about Dawn,**

it was raining pretty hard again while we were walking inside and back out of the store. However, Dawn was the type of woman to appreciate the beauty of the rain. She had no need to hustle through it, and we casually strolled both times through it as people were practically freaking out around us. I really appreciate the type of woman she is. Smoked salmon, rice paper, shrimp, avocado, bell pepper, carrot shavings, cucumber, and a Thai peanut sauce (½ cup of condensed milk, ⅔ can of peanuts, tabasco, blended to perfection). We slept like babies, but only after watching Black Mirror on Netflix. It made us think about the repercussions of a dystopian future, however, falling asleep we came to the conclusion that it (just like the majority of TV) didn't help us grow as people, it only helped to escape from reality, but why would we want to escape?

Day 11 and 12: May 30 and 31, 2018

Dawn was gracious enough to drive us 20 miles out of the city under the condition that we take some of her European chocolate and quiche with us. We loved and appreciated the time we spent with her, learned how to make a few meals, and even made rough plans to come back next summer to really learn how to cook. We walked all day, somewhere from 18 to 20 miles to the next town without so much as a brake tap from any one of the multiple empty bed trucks with only one old white guy driving. That felt like the longest consistent walking day we've had all trip. To be fair, we took a lesser known highway in the hopes that it would be quicker towards Nashville. We cursed the hell out of every driver for the last seven of those miles, calling them every name in the book for not giving us the time of day. We couldn't bring ourselves to give up

1. because our will power and tolerance had already grown so much since the start of this trip that we still had reservations of hope and
2. because we were so consumed by anger in the moment that we couldn't let it go and used it to power us through.

When we got to the next town, amidst the scowls and stare-downs from the entire town, Jake's younger brother, Sam, called us to tell us about his first date with (whom I can only imagine to be) a stunningly beautiful woman, and how he took her off of her school bus and kissed her right in front of everyone. The cajones on that one, **takes confidence to do something like that**. That lifted our spirits a bit and we made it to the mom and pop convenience store down the road where a Santa Claus-esque man by the name of something like Billy Ray gave us local knowledge of a close-by spot where the trains 'go slow enough for a man to jump on.' With vigor in our step, we walked the next couple miles to the place and waited. And, boy, did we wait. At first, about 10 minutes go by and here come two trains going the direction we wanted, but way too fast. Then an hour, and another train comes by in the opposite direction. So, there we were trying to predict and formulate hypotheses about the train schedule. Then another opposite way train comes, then two more.

There was no rhyme or reason to the madness. All night, we waited by the tracks doubting the faith we put in Billy Ray, but how could you not trust a man that looks like Santa? We fell asleep in the rain at 2 a.m., woke up to find a decent place to set up the tent, fell asleep again, then woke up AGAIN to our surprise at 6 a.m., with a train stopped right in front of us, oil carts on one end usually means the train is going the other way, the way we wanted. **Seems practical enough, if the oil carts were near the front of the train and it derailed, it would explode, not to mention destroy the cargo behind it. It's safer to put oil on the end of a train just to be certain that it's least likely to cause any real damage.** We wasted no time finding the perfect train car, it practically had a cage of ladders on all sides around its small platform, perfectly situated under one of the ends of the car, and a small eight-inch wall all the way around that acted as a last resort for not sliding off entirely. We were all ready to go when I heard footsteps and radio talk. I had my arm over the side of the car and was propped up inside quite visibly, so I assumed he saw us. He walked by and kept his head down, too focused on his destination to do his job and observe the world around him. I literally could've flicked his head. But he kept walking as Jake and I continued our deathly silence, which quickly became a bewildered laugh. We hid our bags below the walls and quietly snuck around him as he came again making his checks. Every time the train stops, we do the same routine, thinking we might be inspected again. **Better safe than sorry when it comes to fulfilling a childhood dream.** It's noon now, and the train we chose well missed the turn in the tracks for Nashville a few hours ago, and we have been relatively clueless as to where we are or where we go from here. The train always seems to blow through towns and stop in the most desolate places. We're waiting for a town to hop off into, frustrated that we were thrown deeper into the South, and into the heart of Alabama, the one place we desperately do not want to be. We've seen too much hatred to last us a while. Hatred towards others, themselves, and a blatant apathy for the common man. Jake says we might be heading towards Memphis, which sounds too good to be true, but all that matters is that we're heading west and actually doing something I've wanted to do my entire life.

'Every man dies, but not every man lives' - Jason Aldean

Fucking Santa Claus. We rode that god damn train 120 miles to Muscle Shoals, Alabama. **Back in Chattanooga, we were only 100 miles from Nashville, and now we were 200 miles. Thanks, Santa. One more reason not to believe in you.** Supposedly, the origins of many big artists and labels, FAME Records, Cher, The Allman Brothers, etc. Everyone we talked to in that town made a point to remind us of the rich history there. We jumped off right before the train yard there so that the watchtower wouldn't see us, all the while trying to convince ourselves that we were fine and that we could solve this problem and just move on to the next town. Our water was low from the long hot day. Greg was working for the train yard and followed us for about a mile or two from his truck on the other side of the tracks as we walked through it. He inquired about our purpose for being there, but thanks to my silver tongue, he ended up letting us have a few of his mini water bottles from his cooler, and told us how far town was. **We felt unstoppable walking out of that train yard, even though we were beat from the sun hitting us all day. We wore those water bottles like six shooters in the waist straps of our backpacks, having accomplished another something unlike any other experience. They say you should prepare for up to three days of not stopping when riding a freight train, but honestly, if we didn't stop there, we might not have had enough water to make it three days.** Another yard worker told us which tracks led to Memphis, or west at least, so we debated the practicality of hopping another train, but figured we didn't have the proper resources, so we walked another two miles to the next town, still telling ourselves we were fine. We made it to the Dollar General, **which are about a dime a dozen down in the south and seriously just the fuckin best,** filled our water, got snacks and sign material, and we were feeling good thinking we could make it. Matt picked us up in his dead father's truck not but a quarter mile down the road. **His dad died from lung cancer.** My initial impression of him wasn't the greatest, I doubted his intentions, and honestly didn't trust him. He looked "street", talked slow and quiet, and chain smoked one after another. He told us he'd been doing halfway homes for troubled kids for the last seven years. When we told him our plans, all he said was:

"Well, shit, you're fucked. Nobody gets out of Muscle Shoals, easily, that is."

And at that moment, Jake and I both couldn't hold back the tidal wave of conflicting thoughts and emotions. I'm ashamed of myself for thinking the worst of this man, when really all he wanted to do was help, but that goes to show the change in my mental state and general perspective on people. **Thanks, South.** He told us about the local airport, where we planned to fly out the next afternoon, and even drove us to the other side of town to a really cool handmade wooden walking bridge across the river. We said goodbye, shook hands, and thanked him for his time. In only two weeks, our spirits had been raised and robbed from us simultaneously. Our impression from the South is that the general apathy of the area changes you to the point where your perceptions of life itself are skewed. You are a product of your environment, and if you want to become better, then you need to shape your environment accordingly, so we did just that. **Unless you actively try to change the way you think, whether you want to or not, you will start to become what you surround yourself with. As humans, we are all malleable, but it takes the action of choosing to think independently to become the person you want to be. Mind over matter. Always take life with a grain of salt. Blah blah.** We walked to the airport that night, conveniently on the other side of town again, got swerved at, then woofed at, for our final time in the South. We camped in the outfield of a baseball complex across from a beautiful wheat field, with a large gnarly oak tree positioned square in the middle, ignorant to the fact that damn near every night we've slept in the South, it rained and stormed. We were only a half mile from the airport, but we found comfort in knowing we were leaving soon. **For a place so beautiful, it makes one think that some of the people who live there might not deserve it.**

Day 13: June 1, 2018

What I appreciated most about that field of gold was the various patches of weeds that arose sporadically throughout. It showed that the people there didn't use weed killer, and that they were honest and true about the land. Waking up in the middle of the rainstorm at 2 a.m. was fun. I felt the first few, then heard the thunder, and instantly knew we needed to go into the dugout. Jake took a few pokes to wake up, then called me a lifesaver for pointing at the closest shelter. For anyone who doesn't know me, I really don't like praise, commendation, or really anything that makes me out to be better than I am, which is really nothing in the infinite universe we live in. I also don't like complaining, which it seems damn near everyone is professional at, **unless done ironically about stupid shit and pointing fun at how stupid something is because that's funny 90% of the time.** Jake slept directly on an ant nest and didn't realize until morning when his bag was literally filled with a few hundred. Then again in Nashville that night when his shoes were still full of ants. Funny shit. He's one of the most perceptive people I know, that's why I love him like family, but on rare occasion, awareness isn't always his strongest characteristic. We started the morning off right, singing along to some good ol' country music for a good part of the day until noon, sitting through the last part of the storm. There was a calming beauty to the intensity of the rain. We both have always loved the rain, and taking our time through it even more so. We walked to the airport at noon, talked with a few people there, one was a wise old black man in the corner who put words to the experiences we had felt, and to most of the thoughts that I had. I felt a twinge of pride in myself for at least having thought of his teaching before anyone had told me to. He didn't seem too keen on me though, mostly making eye contact with Jake and whenever I tried to add a comment, he brushed me off. This made me uncomfortable, which further ensued the cycle of me trying to comment so I could be a part of the conversation. Those mosquitoes fucked me over on the baseball field last night, and I counted an easy 40 bites on my right forearm, another 30ish on my left, they even bit my ass and I'm not quite sure how because I was sitting the entire night. So, here I am, looking like a tweaker scratching myself constantly, and the little fucks even gave me a couple of rashes on my legs. To be honest though, I prefer them to most of the people around here. Then a cute little three-year-old girl came over and was all sorts of talkative. Her grandmother's flight was cancelled due to bad weather, and her mother, Nekiar, came to pick her up. All three were amazed at our story, and all three were some of the sweetest and funniest gals we've met on the trip thus far. I also thought the mother was quite beautiful in a strong and confident fashion. She served in the Navy for 15 years, military experience being a prerequisite for just about everyone we talked to down here, which was an honorable characteristic of the South. **Nationalism is blinding, and leads to bias against other countries. But if you do it for a noble reason such as making sure the next person gets home okay or fighting against oppression around the world, then it is a beautiful and enlightening experience to have, one of which I value in people I know.** Since almost everyone in the South had served, when you thank them for their service, they act like that's just how it is, almost as if it's unusual not to serve. We sent them home with half a pizza we had ordered while waiting. Even ordering the pizza was a bitch and a half. The first time we called, I was on the line with a complete and utter dumbass. In their little piece of crap town (sorry, Nekiar, but the only thing that impressed us was your family ♡) with one little piece of shit regional airport, the dominos was a few blocks away. I told the fuck that we were at the airport, which implies that I'm not from the area and don't give a shit about the address, but he still had the nerve to ask after a few minutes on the line,

"So, you don't know the address?"

After the dumpiest fucking pause you could have on the phone. I nicely said,

"We typed in 'Airport' on google maps and it took us here."

BECAUSE THERE'S ONLY ONE AIRPORT. But what does he say?

"So, you don't know the address?"

Again. Like I was just spouting garbage in another language to a machine with only one response. What I really wanted to say was,

"Shut the fuck up, you lazy sack of tots and type seven letters into your hand-held knowledge of fucking everything in the world and bring us the damn pizza."

But that's rude. So, needless to say, I hung up and called again 10 minutes later to talk to someone else. We flew a puddle jumper to Nashville, and after another airport backflip from Jake, a ride on the luggage carousel, and a hella expensive taxi ride to our Motel 6, we settled in for the night across the street from a biker gang bar where they didn't stop revving their engines until never.

249

We washed our clothes at the coin-op laundromat only a block away, and in that hour, I think Jake and I got the most street lessons we've had consecutively. We were in a relatively rough neighborhood to put it likely. Kirk, a deceptively middle-aged skinny white man **(he looked about 29)** who told us all about how he was a 'professional vacationer' **(nice way of saying he lived on the streets and did drugs regularly)** for the last seven years, and how great it was in Nashville for people in his position because the cops only slap your wrists for most things. You could walk around intoxicated and sleep wherever. There were lots of homeless people, both regular Joes and performers alike. When in Nashville! He told us about one time when he was brought to the police station for disorderly conduct, but they never searched his wallet where he had his drugs, then proceeded to do them right after he left holding. Pretty sure the woman he was with was a prostitute, she was waiting outside most of the time, but I really appreciated the way he went about describing her, saying that everyone deserves a chance. He let us know about the type of people living in that area, and said,

"They look at you, as though you're supposed to know what they're thinkin' without saying a word."

Just after that, a shaggy looking tall, thin black dude came in wearing baggy clothes. He had his entire hand down the front of his pants (seemed to be legitimately holding a weapon) and stopped in the doorway, door still wide open, staring at me. At this point, Jake and I were doing a little home-workout in the laundromat to take our minds off of things and let out some pent-up aggression. We were both shirtless because both of our one shirts were gross and being washed, Jake was shadow boxing himself in the reflection of the dark window, and I just finished doing daily push-ups, so I had a sick arm pump (please know I say 'sick' ironically), smoking my e-cig, looking relatively stern. For a long moment, we stared at each other without either blinking. I wasn't about to deal with any shit, and he saw it in my face. In that instance, I knew what Kirk meant and I saw the intent through the intensity of that man's stare. I gave him a head nod and he walked out without saying a word. It didn't help the tough persona I was trying to present when Jake was doing cartwheels and running across the street goofy for the rest of the night. When we were talking with Kirk, he let it slip how much money we had saved, and even though Kirk seemed somewhat trustworthy, I was livid. **We had a conversation after and he realized it was a mistake, another thing I love about Jake is that he's able to admit when he's wrong and is usually pretty good with logical conversations about the nature of things.** It was understandable that he was excited about being out of the deep South because I had the exact same feelings, but he was calling a lot of attention to us in a place where attention was the last thing you need. He commented that he likes serious Joey as much as I liked focused Jake, but I had to remind him, perhaps even teach him, since people don't really think about it, that there's a difference between being focused and trying not to die, and that I felt it prudent to exercise caution in an environment that demands it, not on the road.

Day 14: June 2, 2018

    We hit the strip today, walking up and down Broadway, going into Jason Aldean's bar, the Tequila Cowboy, Florida Georgia Line Bar (where I had the most amazing fish tacos I've ever had, creamy spicy sauce, lettuce, salsa, and sweet corn). We went into Wanna-B's karaoke bar and sang Red Solo Cup and got the entire bar fired up. We saw a performer who re-did Hozier's *'Take Me to Church'* on the street, and damn did Greg put some stank on it and really make it his own. We also saw Stevenson Everett in the Tequila Cowboy while he was singing my favorite country song of all time when we walked in, Garth Brooks's *'The Thunder Rolls.'* In our final bar, I talked with the bartender gal who also happened to major in astrophysics, so I told her to message me sometime and we could talk about physics, super sweet gal. We walked the bridge that many tourists seemed to flock to. It overlooked the city and was quite a sight as the sun was setting. We considered jumping onto a ferry boat passing underneath and seeing where it took us, but not even 10 minutes later, it made a U-turn in the river. It would've been a short-lived trip for a painful 25-foot drop. Jake and I discussed the down sides of being as perceptive as we are. In my experience, I find it easy to influence people, talk to them, and all that jazz. So, when I see a beautiful woman, it's not a question of how I can take her home, it's about why would I. I think initially about the romantic ways I could make her happy, which could mean a lot of things, not strictly sex, but diners, dates, you name it. But then I can't help but think of the other women I've fallen for in my life (the latest of which has been ignoring me for the majority of the time I've known her, funny enough, **and even when I write this, I find myself in the same situation with another stunning woman, someone who I think I'd be strong enough to commit to**), and how there might be a guy out there somewhere who feels for that particular woman the same way I've felt. Then I think about how she's someone's daughter, having a good time with her friends, and all these other factors that lead me to think,

    Who am I to interfere with her life? How selfish would I have to be to become another thing in this woman's life to think about or worry about? I should just let her enjoy her life.

It's very easy to hit on women, **or at least approach women since I can't attest to how much they actually care about what I have to say,** but why should we as men have the inert desire to complicate a woman's life? Like when Jake and I sang karaoke, we could've just as easily stayed and flirted it up with the bachelorette party ladies, even when a few came up to us and were talking about how good we were, but we both just thanked them and left. We let them keep having a good time, and got on with our lives as they were. It sucks, in a way, being as perspective as we are, **not bragging, it's just a lot of thoughts to have to deal with on a daily basis,** and often growing up I wished I could have lived in blissful ignorance, but what kind of life is that? **Post-trip, I truly feel more confident in myself for the way I think about the world, and I'm glad I have the capability to think for myself so much so. I only hope more people can grow to become perceptive, because thinking leads to imagination, and imagination is the truest form of individuality.** Easy? I don't like easy anything. We walked back to our motel after all the bars turned 21+, it really sucks not being 21; **I'm 19 if you can believe it,** and we got a solid five hours of sleep. Almost forgot about Willy Reese. Usually, I don't indulge in talking to many homeless people, not because I don't like them, but there's nothing I can do for them, being a broke college kid myself, but truthfully quite a few of them never seem 100% sincere **(probably from being rejected by society their entire lives),** but I try to look with understanding and make sure they feel heard because it sucks being passed by without so much as a thought (thank you, hitchhiking). But at least, I acknowledge them and give them the,

'Sorry, but I don't carry change with me' excuse.

Jake almost seems too preoccupied with other thoughts to give them the time of day. He's not selfish, but doesn't easily trust the people who he doesn't feel can benefit him either with conversation or help him with his life goals.

"Most people end up disappointing you, so there's not much point trying."
– Jake

I know we tend to blanket statement people as 'most people' or people in general, but we're not talking out of our ass. It's just an observation formed by people like those we've interacted with or seen throughout our lives. In our experience, most people do or don't do the things we talk about when we write blanket statements, in turn, Jake's input is derived from his experiences and is a perfectly reasonable conclusion. So, Jake brushed past Willy, but I could see something different in his eyes than in most homeless I've talked with. Of course, he started the conversation with what he needed, which was $1.70 for transit fees to a homeless shelter, but I wasn't all too convinced, so I told him I'd buy him lunch. He said he was in a real hurry, and that he would even perform or do anything just to get bus fare. **That, in itself, is disheartening, a full-grown man having to belittle himself just to gain sympathy from an apathetic world. Makes me sick to my stomach how we treat each other, but I know I'm a product of my own environment too, so I have no moral high ground to speak from.** I told him to walk with me and tell me his story since Jake and I were walking to the Florida Georgia Line Bar, where I considered getting a few bucks for him. There was an air of solemnity about him, his eyes weathered by the southern heat and disappointment, and his face was that of a sulken spirit; seemingly hollow behind the mask he put on. The only aspect of livelihood about him was the fullness of his nose, which even too was scarred along with his shoulders and neck. He said his wife had left him recently and he was meeting with a friend about possibly living with him before going to the shelter. I still wasn't entirely convinced, but his serious eyes and his desire to work for it made me feel like it was the right thing to do. He had just walked from the previous park a few blocks away downtown, and told me how he asked a couple older ladies to spare a little change, but they treated him with such disrespect saying,

'Get a job, you filthy nigger.'

One of their husbands also pushed him hard with his cane, instead of simply moving on as a group, they went out of their way to make sure he felt disrespected and worth less than their precious lives. The ATM only dispensed 20's, so I gave him $20 and when I handed it to him, he was taken aback, astounded, in disbelief that some random kid could be so generous. The tears in his eyes made it all worth it **(not in a weird fucked up way, but there's a fistful of emotions associated with making someone's day, and possibly restoring a little of their faith in humanity amidst it all),** and I knew at that moment I had done right. That's a feeling I live for, and I would wish it upon anyone capable of seeing the importance of such an act. Also, an old lady asked if I worked at the country music hall of fame because of the hat. **Oh yeah, we also went to the Country Music Hall of Fame, it was pretty neat.**

Day 15 and 16: June 3 and 4, 2018

Fucking Greyhound.

We left our motel at 6:30 a.m., walked to the greyhound station on the other side of town, and got there at 8, with our bus leaving at 9:20. The bus didn't end up leaving until 10, but the driver was the most seasoned, salty motherfucker, who had worked seven days a week for the last 30 years as a driver. So damn funny too, he would call people by their defining characteristics, like one lady he called 'Big Girl' because she was a little larger than average. Another, he called 'Camo' because he wore a camo baseball hat. **It might not sound funny on paper, but the way he was ripping into them was pretty damn funny.** Boarding the first bus, there were no empty pairs of seats. At this point in the greyhound experience, I'd noticed that a lot of people had either looked at our hats and giggled, or even a couple of them cat-called to us 'Crikey!' in reference to Crocodile dundee, like they thought they could phase us. **Bitch please, we don't get phased anymore.** But you just gotta roll with the jabs. I have pretty decent hearing in most situations, so when we got to St. Louis, it wasn't really too much of a surprise that I heard more people making fun of us, but I just smiled at them because I love looking like Indiana Jones, he's a badass, and fuck what they think. When we took our seats in Nashville, I asked my seatmate what his name was, he said he didn't have one, and that I was a stranger, and that he didn't talk to strangers. Okay then. What a way to go about life. Good ol' No Name also went on to talk about government conspiracies with some 16ish-year-old gal that looked uncomfortable around him 98% of the time, but somehow he knew her, introduced himself, the whole nine yards, even though they didn't get off at the same stop. The gal initially sat next to a man, Jason, who took up about 1.5 seats, and she was complaining to No Name about her seating arrangement. No Name said, right in front of Jason, I'll have you know, and quite audibly,

"I told you, you should've said no to that fat man, there's not enough room for you two,"

I was in the aisle seat of the bus, he was by the window, and she was in the aisle seat just across the walkway, so they were practically talking right through me. Clearly fat shaming him, and need I remind you, RIGHT

IN FRONT OF HIM. I looked at No Name with a stern face for a good solid few seconds, then offered to switch seats without breaking eye contact, mostly to shut him the fuck up and hopefully make Jason feel a little better. He had a truck driving job all lined up in Kansas City. Seven hours later, we got into St. Louis, then from there directly to Kansas City. An hour layover in Kansas City later, we left at midnight on the direct bus to Denver. Initially, Jake had passed the only pair of empty seats and was making his way to an empty one next to some random guy, but I made sure to say,

"Jake, what the hell, man? I didn't want to sit by you anyway…"

We both had a chuckle and sat together. That goofy motherfucker. The older black man next to me, **who had initially sat next to Jake on the very first bus,** was going to San Francisco, and was practically illegible, save the multiple times he tried selling us both weed. Jake finally got the chance to sit with me after the layover, where we spent the next 10 hours miserably drifting in and out of sleep, listening to the guy in front of us go off about how he was in a cartel and how he fucked lots of bitches, even in rare cases fucking the daughter, mother, and grandmother of a family. He took no shame in talking about the married women he'd slept with, or getting his dick sucked in church. **There was even a time when he fucked a lady, then smoked pot in her child's room while the family was out.** His seatmate, Aaron, was surprisingly insightful for his level of street. He was trying to get Cartel Guy to think about his karma, energy, and impact on the world. To be fair, Cartel Guy wasn't all that bad of a guy, it was just easy to see that he had fallen into the classic rut of what society and social media dictate as a 'cool' life, one that people are brainwashed into believing they want. **It's an ever-perpetuating cycle, where the idols of our generation promote a falsified reality through their social media. That in turn inspires the next person to want to live in vanity and for unjust reasons of sex and money. They then continue the cycle by talking about their 'cool' lives and posting all over their social media, which then inspires the next wave and so on.** He had tattoos all up and down, said he dealt drugs, fucked bitches, and by society's standards, he was a well-made gangster man. He even talked street, although his eyes showed a remorseful tenderness of a man trying to escape his own reality. **Isn't that what we're all trying to do? Find a way to escape the things we hate about ourselves? I know I do, and when I look around, I can see it in others. Not everybody hates themselves, but nobody should have to escape themselves either. Gotta learn to roll with the punches and accept yourself for who you are, but that doesn't mean you can't work to change yourself. It's better to at least do something than nothing at all. How else do you grow?** He was driving to meet and live with a woman he met online. Assuring us she was a 9.5/10, and that he was going to fuck for housing. He was only escaping. Upon arriving, we immediately brushed our teeth and cleaned up after being given the privilege to after 36 hours of doing nothing. We called an Uber to get out of the middle of the city, considering it was two o'clock and we wanted to make it to the mountains (40 miles away) by nightfall. Our driver, Josephine Marie, was an MMA fighter of sorts, having trained to defend herself from a crazy ex-best friend who threatened to kill her after she wouldn't date him. **Most days, I feel ashamed to be a guy because of the fuckin batshit crazy actions of every other guy I hear about. A lot of guys don't have the capacity to get past their own emotional barriers, or even consider other people's perspectives before they act. That's why I value women well above any man.** She was a beautiful gal, not an immediate beauty, but the type that grew over time and as a result from her positivity and general disposition on life (she actually recommended the place I'm writing this at now, two days later, but I'll get back to that in due course). Jake ended up leaving his hat in her car, though luckily enough she gave us her business card for her real job of being a personal trainer and nutritionist and all that good stuff. That was lucky because Uber doesn't give out phone numbers, so there was no other way to contact her, so as Jake was already counting his losses, I gave her a call and she hustled right back. Great woman, to say the least. Then we walked through Boulder. I was a bit cranky since I barely slept on the uncomfortably smelly bus and from dealing with a lot of relatively shitty people even if the majority I hadn't met were more than likely decent enough. But I wanted to just get out of the city and back into our groove, and in those few hours I had turned into focused Jake. I was still making efforts to appreciate the art and beauty of the Rocky Mountains surrounding Boulder, but Jake was in an entirely different state of mind. I wanted to punch him a few times, I swear. He wanted to tour the campus **(in retrospect, I wish we had done because even from the little bit we did see, it was gorgeous),** walk inside a neat looking apartment complex, go into every café, make conversation with people sitting on their porch, and on the street, he was really taking his sweet time, but it was late afternoon and I wanted to make sure we had a good place to sleep. We walked through a park on the last stretch before the highway that led out of town. I indulged Jake's desire to take a break; we were both tired, and he went to go dip his head in the adjacent river, and I did this for two reasons,

1. I thought he might settle down after running around a bit
2. A little alone time would be healthy for our friendship

3. In the end, it did help me calm down and helped me to fully appreciate my surroundings, and the beautiful simplicity of the lives of the people enjoying the park with their friends and family.

Life doesn't have to be complicated. We take it upon ourselves to make our own complications or our own joy. About an hour passed and Jake came back after doing yoga with more beautiful college women and running around a bit, like I originally thought he would. We set off back down the highway with no rides for the next three hours, but more appreciative to be marooned in such a vastly different and extremely enchantful place. **Not to mention, a limitless supply of fresh water coming through the river we were walking along. Mountain water is the freshest and best water you will ever have in your life. The constant motion of a fast river acts as a filter as the water runs through the rocks to separate impurities, but saves the good impurities that help balance daily mineral doses. Never ever ever ever drink stagnant water.** About an hour before the sun set, and six miles from the closest town, J.P. picked us up. In all honesty, I don't think I would've been friends with him if we met under regular conditions. He kept clearing his throat every five seconds, his car was more than a bit disheveled, but he was coming back from fly-fishing, which is something I like to do with my dad. Originally from California, he moved with his girlfriend up to Colorado because she was going to school in Boulder. Although he was generous, and hospitable, he seemed like the type of person who really placed an emphasis on how people thought of him, almost afraid to be rejected or go outside of his comfort zone to avoid criticism or ridicule. He dropped us off in Nederland, at the town market. The town was full of an interesting sort of characters, very private, wary of strangers, and a lot of the people I saw walking into the store were dressed like 'hick-thugs,' so you can only imagine their personalities, **because I sure wasn't about to go out of my way to talk to them.** Another shy, beautiful redhead gal working inside the store gave us directions out of town. That's another thing Jake and I are coming to find is that no matter what town we were in or where we were, there were always two things you could count on

1. Beauty. There are beautiful women, art, nature, and sights everywhere.
2. Death. Dead animals are damn near everywhere you look, keeping the cycle of life alive and well. **Death -> Maggots-Flies/Soil Nutrients -> Small animals eat bugs/New vegetation grows from the soil for herbivores -> Big animals eat small animals/We eat herbivores**

(deer) -> Big animals keep population in check -> Humankind keeps big animals in check -> We die.

With everything, there is always a duality. Like I said earlier, I strive for duality in every aspect of my life, but the most prominent duality of this trip was that of beauty and death. Within death, there is a solemn beauty in the renewal of life thereafter, where with beauty there will always be the destruction thereof by those who would otherwise not appreciate it. Why are flowers beautiful? In their uniqueness, they do not last forever. They grow, they die, then they are regrown ever slightly different. Somewhat comparable to the saying, 'distance makes the heart grow fonder,' in that when you're constantly witness to something great, its value depreciates over time and it becomes hard to remember why it was significant in the first place. Jake and I have grown up in a truly incredible and beautiful region of the country, and yet I find it hard to truly appreciate all it has to offer because it's all I know. I have no true reference of comparison, and for that, it isn't unique to me. It just is. Death and beauty, one can always be found within the other to some degree, but never can you have one without the other. We followed the highway to the top of the valley, where we had just enough time to pull out our hammocks and comfortably get ready for bed, looking out over the valley and the Rockies.

Day 17: June 5, 2018

That next morning began a very interesting day, and as a conclusion of that day, both Jake and I came to the consensus that the South fucking sucks, and that people are amazing in Colorado. We also had our first, and hopefully last, taste of interstate hitchhiking. Initially, we set off for the top of the mountain without our bags, and even five minutes in, I was getting winded due to the high elevation compared to the last two weeks of hitching. Actual hiking is a whole 'nother ball game. We saw the valley in all its glory just as the sun was cascading across for the start of the morning. On the way down, we saw a man-made bunker long since abandoned, high walls of large rock and a heap of dirt around almost made it invisible if you were looking up from the base. The rocks must've been taken from somewhere else and brought there because they didn't match the area around it, either way, definitely a lot of hours spent making only to be abandoned. It was some 'The Hills Have Eyes' shit though, tell you what. We were only on the road for about an hour when a bus stopped for us with a printed sign on the back, 'peaceful rides.' Jake and I were taken aback at how attentive people already were here, and how much more positivity there was towards hitchhikers in the west, even the previous day had gotten a lot of fun honks, cheers, and thumbs ups. Chip the driver and his pup, Gandalf, were on their way to work. He had been mayor of a smaller town in the area in previous years, taking a hard stance on increasing their economy from $140,000 in debt to $120,000 in the clear. Since then, he drove his 'peaceful rides' bus to pick up disorderly and drunken members of society. For a small fee, of course. **He was an interesting man. First impression was an older hippie, but he was carrying a load of lumber in the bus too, so he was either a craftsman or a carpenter. The inside of his bus was very comfortable, covered ceiling to floor in stickers, and his dog took liberty of walking back and forth between us, loving the attention.** He took us about 15 miles up the road and even then, it wasn't but a mile more until another man picked us up on his way to Central City to drop his daughter off at school. He never introduced himself to us, but from Jake's account he was very enlightening and shared wise words. I sat in the back with the little one, and damn could she talk until the cows came home. About everything under the sun, all the while never breaking eye contact with me for the next 20 minutes. Of course, I indulged her in her little rants about her sister and cousins, and her new sled, and the drama at her school and how she wanted a pony, loved Belle from *Beauty and the Beast*, and was convinced of her personal royalty. She dressed to impress, and wasn't afraid to speak her mind with her own tones of sass and attitude. There's a lot of money in Colorado, and of every town we passed through, gorgeous homes watch the valleys with such unique design and masonry that each home is more beautiful than the last. Central City seemed to be where all the money of the state filtered through, with a minimum of five casinos within a half mile of one another and hotels built into the mountainside. It was interesting to see how so much money can be consumed in such a small town, and the town itself was beautiful just like the last and just like the next. Just about as soon as the town started, it ended, and we found ourselves walking out of town only minutes later. Not even another mile down the road, a rock-climbing couple stopped on their way to a climbing spot to meet up with their buddy and spend all day climbing. Cole and Kelsey shared our distaste for the Missouri/Indiana/Kansas states, given that they both grew up there. They even shared their fruit with us, very sweet couple. It was only 10 a.m. and we reached the most amount of rides in one day that we had ever gotten in the South. Jake and I are in love with Colorado. Lucky for us too though, they said that they never take the route that they had just taken that day, and they took us the rest of the way down the highway we planned on going, then up another smaller/lesser known highway we planned to go down. We then made our way the next six miles to the interstate, where we sat under the main stretch and on the side of an on-ramp that passed underneath. We were there for about an hour with our signs and it was nice to be out of the sun. We just sat, waiting, lighting stuff on fire, seeing how high we could throw branches (trying to clear them over a light post that stood taller than literally any other light post I've ever seen ever. Unnecessarily tall, probably 60-70 feet), just enjoying the day and each other's company. The worst part about interstate hitchhiking is the wait. You're putting all your eggs in one basket, relying solely on the generosity of the few as hundreds pass by. I'm an impatient guy, I don't like waiting and neither does Jake, because you can't do anything else but wait. It's illegal to hitchhike on an active interstate, not to mention dangerous too. So, again, we said fuck it, and climbed up an eight-foot rock wall with our packs to a wide shoulder of the interstate just above where we were waiting. We figured the more face time we get, the higher chances we would have. Interstate vs. state hitchhiking is very comparable to the lottery system, where state highways are the dollar scratch tickets that have a higher chance of winning with minimal payout, and interstate hitchhiking is like the Powerball, with increasingly minimal chances, however, the payout is so huge you'd be stupid enough not to at least try. Only a half hour went by before Joan picked us up on her way to moving in with her mom. Joan was in her mid-50's, two kids, no husband (maybe divorced but the way she avoided it led me to believe he died fairly recently). She really just wanted someone to talk to, however, I've been in a pretty anti-social mood lately, so luckily, Jake bit the bullet on that one for me. She was a bit disorganized too, **as it seemed most people are oft to be,** but understandably. The way she was talking about her mom, along with her own age, led one to the logical assumption that her mom was dying too, so her life was all sorts of uprooted. Her kids didn't visit her much either, which is disheartening because although simple, she's a sweet woman, which leads me to wonder if their childhood was rough because of their father, and the passiveness from their mother created resentment from them. When she dropped us off at 2, we were already 100 miles into the day, and only 30 miles from Hanging Lake (the place where Josephine Marie recommended we go). So, yeah, we were on cloud nine, thinking we already burnt out our luck, but might as well try to get there before the sun goes down. **She dropped us off at a Whole Foods at the edge of her town and there we refilled our food, relaxed, smoked, and just took in the most of an overcast day. It was funny looking around because most of the mountains still had snow in the beginning of June, and this was the first day we started to doubt the durability and warmth of our attire since we really only planned on being in the heat. It was summer after all, it should've been hot.** After a puddle jump ride from a hardcore ladies' man, Mark (who was coming home from his construction job to have drinks with some woman and who told us all about his upcoming trip to South America where he planned to spend six months hiking, only after he spends a month with another woman at her home on the east coast of South America), we took a dive in the 'lake' in Avon, CO. It was such a cute little town, with fancy street signs pointing at little markets and shops around town, positioned right beneath ANOTHER ski mountain of the plethora we've already passed, more beautiful women, and even a few people who walked up to us curious about who were and telling us how impressed they were with our story and our spirit. We took a bus even closer to Hanging Lake, just the next town over for $4, now only eight miles away and at the start of Glenwood Canyon in Colorado. We chatted with the older ladies there, and followed the bus driver's directions to the walking path that followed the entirety of Glenwood Canyon. I was motivated to get us to the lake and camp before we fell asleep, supposedly there was a gigantic waterfall only a mile hike up a mountain along the canyon and that sounded amazing to sleep next to. I pushed Jake to walk faster than we ever have, but with the sun going down and our bodies exhausted, we barely made it to the Bair Ranch rest area along the path. Now only three miles away from the lake, I knew this was the perfect place to sleep for the night, and I am damn happy that we didn't try to keep going, cuz it was a bitch of a hike straight up a mountain with a 50 lbs. pack. The rest area had bathrooms, vending machines, and wall outlets. It was positioned along the river and across was Bair Ranch, a wilderness retreat which looked super cheesy like it was designed to suck in rich people who wanted a 'farm' experience. After charging our phones and eating, we fell asleep on the well-maintained park side of the rest stop. We talked about our final wishes, while I was pushing us to walk faster, if the other didn't finish the trip.

Jake's Final Request: He wanted me to give MS a carrot for her insatiable love of carrots, and his of her.

My Final Request: I wanted someone to tell EMSRP that I wish I could give her a hug one last time.

I used their initials because I don't want to call anyone out, but I hope EMSRP knows I appreciate her as a woman. I try to also live without regret in knowing my loved ones know my appreciation of them.

Day 18: June 6, 2018

Only to wake up at 5:30 in the morning showers of the prompt sprinkler system, which Jake assured me 30 minutes earlier they wouldn't turn on next to us. We spent the remainder of the early morning trying to warm up and dry off inside the heated restroom facility because we would have easily frozen in the Rocky Mountain early morning air. And damn, does it get cold. There's a beauty in that as well, the livelihood of the area experiences the extremes of nature and still holds strong against all odds. The river-sculpted walls of the canyon were astounding as well to say the least. An Amish community rode past us on their bicycles while we were walking to the lake, which reminded Jake of when my aunt discussed with us the foundation of her religious views in the household. They practice Mormonism. We are honest men, and in that honesty, some might find us brash, but we hold steadfast to our outlook of the world and our conveyance of our thoughts direct from it. She told us,

"I may be the breadwinner, but at the end of the day, my husband is the head of our home, but we are equal."

The problem that Jake had with this phrasing was when she said 'but we are equal,' instead of 'so we are…' or even 'and we are…'

Almost as if she wasn't only trying to convince us of her truth, but was also trying to convince herself. How we see it, them being equal means that she has a voice and her opinion is heard, but at the end of the day, if the husband is the head of the house and his decision is final, then her opinion really means jack shit. That's nothing to attest to the character of them both, they're both amazing people with very upstanding and redeeming qualities. But in the belief that she, and for that matter, a few women we know from back home and what we imagine to be all across the country, are committing themselves to this belief, the idea that they are truly okay with being treated as lesser based on their religion (even though they are 'equal') is disheartening. In that so many women are excited for it, look forward to it their entire lives, and live only for that belief. It isn't equality. It's barely even life if your opinion carries no real weight. **We were both raised in matriarchal households with the philosophy that women must be respected, if not even treated better than men, and we truly detest any thought or action against that belief.** We made the hike up to Hanging Lake, which was a mile damn near straight up the mountain, what a bitch of a hike, and Jake did it all in one go. I took about four breaks, but even exhausted, it is so breathtaking. It's hard to find words to describe the beauty of this clear lake as well as the magnitude of the waterfall above, which I calculated as going 35 mph. We've picked up a lot of trash around here, which really angers us both at how careless and easy it would be for so many people to ruin a work of beauty. Watching the droplets fall from the veil covering the backside of the lake, I thought about time and life. If you talk to older people, they say time only moves faster and faster, and the more you grow up, we start to see our 'time' accelerate more and more. It starts slow, gaining speed on the way down to the collective of lost souls and the timelines of those before us, merging into one, becoming part of a whole we know as history. In that sense, time is almost like gravity, or at least the flow of water which works in the same principles, and the faster you relent to the flow of an uncaring and callous rapid of eternity, the faster you return to the natural state of lower potential (death) that we are all destined towards. The universe always prefers the state of lowest potential and the least amount of effort possible, all moving towards an energy state of zero or close to, so then it would only make sense that life, in theory, shouldn't exist in the first place, or that we're destined as a race to never exist at all.

But only if we give in.

We must fight with every ounce to never give in, not only for ourselves, but for our species, or else all is worth not but meaningless specs of dust, or just another drop in the lake. We finished walking Glenwood Canyon, carved by at least a million years of erosion by the Colorado River, and comparable only to the intensity and sense of awe provided by that of the Grand Canyon, through which the same river flows. The canyon itself spans 13 miles, with an interstate blown right through the middle of it all. Weaving in and out of the mountain, it truly is a feat of man and the brainchild of President Eisenhower when he decided that he wanted our roads to be one with the land. The Colorado River itself also provides 10% of the nation's river-water supply, creates the Arizona/California border, and runs clear into Mexico. We left Hanging Falls around 2, and made it to Glenwood Springs around 7, with just enough time to get in a sit-down dinner at the Ranch House restaurant, and pitch the tent in the closest park before it got completely dark. Rolled my ankle a little running downhill from Hanging Lake, so that's inconvenient, but it doesn't hurt all that much. **I have a very high pain tolerance, but I'm relatively in touch with my senses, so I know painful sensations exist, but the association of 'hurting' doesn't really exist in my head. Except for stupid little shit where there's a lot of force concentrated on a small**

surface area of skin, like stubbing your toe, that hurts like a mofo. Or 8+ out of 10 pain levels. Also had my first elk burger, and now I realize why everyone who's had it goes crazy over it. It's pretty damn lean and juicy, really absorbing the flavor of the fire, and just barely gamey enough to give it its own unique flavor as well. Which reminds me of a paraphrased quote from Marco Polo, the 13th century explorer,

'I see not beauty, for a woman is just as beautiful as her flavor.'

Which I perceive to mean that beauty is never strictly on the outside, but rather the taste of her character, which permeates thereafter.

Day 19: June 7, 2018

As we woke up the next morning, we made our way to the 'City Market,' which is a brilliant decision to name one's corporation something that sounds local, but as we later realized they were in damn near every Colorado town we went through. The manager tried to kick me off the property while Jake was buying his food for the next few days. **We do our shopping in shifts so that there was always someone watching the bags.** He did this because he thought we were begging, which I find extremely insulting to my character because I've never begged for money in my life, nor will my pride ever permit me to do so, but he never stopped himself to ask what our intentions might be, only assuming the worst. **I was polite enough to explain that we were actually shopping at his establishment and that we weren't begging, apologizing for the confusion. Jake and I have a general distaste for apologizing, because we don't act unless we truly feel what we act upon. Based on this introspective, we can only assume that the majority of apologies come from a place of apathy and insincerity. For me, however, apologizing is more for the other person than it is for me, other people want to feel heard and appreciated, so even if my apologies aren't always sincere, at least the people I apologize to have closure on the issue.** Doc Holliday's grave was but a mile up the road, so naturally I'd want to see the resting place of literally the fastest gun in the Old West. Whiskey bottles, playing cards aligned in royal flushes, and dolla bills upon dolla bills. The tributes of those who passed before, honoring one of the original outlaws turned vigilante, who paved the way for the mentality of living in the moment and seizing each day for everything it had to offer, even until his final days. His famous last words were, 'This is funny.'

That being killed by tuberculosis rather than by bullet. A short walk over the interstate brought us to the 'nation's only underground vapor caves,' or at least that was their gimmick. I tried smoking my vapor cigarette down there **(I'm an addict, but more in the sense that I have nothing else to do with myself, so I do that. Life is generally boring until you make something of it during the majority of the time spent waiting for the next exciting thing),** but the mineral vapor got sucked into the filter, so I was basically smoking mineral water, which has to be good for you, right? We stayed there until about noon, where we jumped back on the side of the interstate just before James invited us in his truck and offered us some of his weed, him being a hemp grower as well as entrepreneur, and father of five boys and one girl. **I remember walking out of town to the interstate, we filled up our water one last time at the last gas station as we had learned to do, and there we struck up a conversation with a woman who was passing through. She was a gorgeous woman, not only because of her straightforward attitude, but one of those faces you would marry if she ever asked, just to try it out for a little while. She wasn't wearing a bra either, and I know what you might think, 'ignorant college boys only think about boobs and sex,' however, it was the way she carried herself with no shame or embarrassment for who she was, rather proud of who she was with no preset anxiety of what other people might think of her. She offered us money, we declined, then she told us if she saw us on the interstate, she would give us a ride. She also mentioned that she would be coming through later in the day, so if we needed to stay with her, we could. Every immature fiber in my body was telling me to wait on that side of the road all day to wait for her, and hell did I want to. I held in my frustration when James picked us up, but just as soon as it came, my anger was erased by the gratitude I had for the hospitality of our driver.** Very religious man who felt it his duty, nay the duty of humanity, to act with kindness, to invite all others into his life, and help to the best of his abilities, just as Jesus would. Probably the most Christian-esque qualities we'd met thus far on the trip were embodied in this man. It turned out he was also a carpenter earlier in life, though you could tell from his slowed speech and minimally extensive diction that the weed had taken a toll. He dropped us off at the next 'town' 15 miles away, and naturally we jumped back on the interstate. We didn't have to refill our water, and we were still ready to go, so we said to hell with it. Little did we know that that particular section of road had seen the highest fatality rate in the entirety of Colorado, only having known so after David the local sheriff had received calls from other passing officers concerning two crazy people hitchhiking on the side of the road. Eventually, he came down to tell us so right before giving us a $20 bus card to get to the next town of Rifle, CO. He would've driven us himself, but he had a sergeant's exam and interview soon. After we got into Rifle, we thought we should at least get out of the city before sun down, which was an hour and a half away, tops. **Do you know how to tell how much time is left in the day using your hand? While standing at ground level, if you hold your hand at arm's length then align the bottom of your pinkie with the horizon of the ground, every finger is the equivalent of 15 minutes of sunlight** roughly. **Then you work your way up to the sun and count how many hands it takes before the sun goes down. If you have three hands until sundown, then that's three hours of the day you have left.** As we approached city limits, Doug and Julia **(to be honest, I don't know remember if his name actually was Doug, but I know Julia was her name),** an older and well-connected couple from the next town of Meeker pulled over in front of us and asked,

"Do you boys want to go on an adventure?"

Us, being of the adventurous spirit and without a breath wasted, agreed, and they began driving us the scenic way to Meeker. Julia herself being raised there with some ungodly knowledge of the schedules of damn near every person in the entire town of 2,500, and Doug having met her as the town diesel mechanic some time ago, seamlessly moving into the regional director for the natural gas operations, drove us down Old Highway 13 a.k.a. Flagg Creek a.k.a. the original wagon trail through northwest Colorado. Julia was hesitant to open up to us at first, even giving Doug a questioning look as he offered us the ride, but as soon as we got her talking, the flood gates opened and they treated us with the same graciousness they would towards their own grandchildren. Doug was a simple man, but made well of himself, driving a brand new 2018 Jeep, knowing the slightest distinctions between the 50 head of elk we saw in a single field, pointing out two brown bears that he knew were in reality blonde, black bears based on his knowledge of physical anatomy, concluding with interesting thoughts on the ozone layer. Although both of them believed in global warming, neither thought it the product of human consumption. Julia had worked in park service for 10 years, having shared some of her stories with us the most memorable one in my opinion was the time when she was asked:

"At what elevation do deer turn into elk?"

That was fucking hilarious. They gave us a tour of the town, finally dropping us off at the park where we paid for our first, **and only,** night of sleep. John and Lisa were in between apartments and camping in their trailer in the parking lot across the small yet powerful river from where we pitched the tent. John was working in town and all, and Lisa was his wife who passed through every week or so to keep an eye on him. Through Jake's efforts of striking up a conversation with them, they offered us dinner and we swapped stories of our travel and their military experiences. Never before had hot dogs and chili atop a hamburger bun tasted so good. They also had their fair share of children, three boys and two girls. Treating us as one of their own, they made sure we had plenty to eat before retiring to their trailer for the night.

Day 20: June 8, 2018

Damn, I wish I was 21 on this trip. We're camping just outside of a beer festival in Lander, Wyoming tonight and it sure looks and smells fun as hell. Although, to be honest, I probably wouldn't have much of my already low spending budget afterwards.

Daily Thoughts:

Google definition of **Photon**: A particle representing a quantum of light or other electromagnetic radiation. A photon carries energy proportional to the radiation frequency but has zero rest mass. **This means photons are proportional to wavelength as well.**

Photons expand and contract based on the wavelength present. Because photons have to transfer all of their energy simultaneously, the capability to expand and contract is reduced since every wavelength occurs at the same time through light. Due to a higher energy and a higher frequency, the wavelength is reduced and photal contraction would happen. Due to lower energy and lower frequency, the wavelength is elongated and photal expansion would happen. Because contraction and expansion would happen simultaneously, it creates the size of a photon that we record and measure in labs and light tests.

Different seasons/hemispheres: Because the northern hemisphere is tilted towards the sun during its summer months, we experience a higher amount of heat/energy (those words are very interchangeable in my vocabulary). Given that higher energy contracts photons, being closer yields a higher heat since the higher energy has less distance to travel whereas the southern hemisphere is further away, in that the longer wavelengths of lower energy reach more so.

Cloudy Days: As the sun initially comes up on a cloudy day, it seems more often than not, the furthest part of the sky from the rising sun seems to dissipate its clouds earlier than the rest, possibly since the lower-energy/longer-wavelengths of light apply the most direct influence over the atoms in the clouds. People can often get sunburned on cloudy days, where typically on a normal day the higher energy (UV rays) are somewhat shielded by lower-energy wavelengths that counteract photon contraction, that shielding is lost when the lower energy interacts with the atoms in the clouds, leaving the ultraviolet energy free to plow straight through and affect us.

We woke this morning to Lisa's call for bacon, eggs, and English muffins, topped with her homemade Colorado peach jam, comparable even to the peaches of Georgia, yet missing the subtle flavor of racism we had grown to know. They offered us cash to help move John's heavy belongings into a 'new' apartment, but after their hospitality, we figured it was the least we could do to help and negotiated to do it for free, and if they gave us a ride, that would be great, but we expected no money and declined when Lisa offered us. **Turns out, we were only moving from a second story apartment to an apartment on the first floor.** I revealed to her that Jake and I didn't necessarily believe in a god, her being a devoted Christian through and through, and had even spent two weeks in Ethiopia teaching/preaching Christian views. She had us pray with her before we left, to an extent that seemed forced and impractical, asking god to

"please reveal yourself to these two young men."

As a man of science, while we were talking, I respectfully shared with her that I didn't find the idea of God practical, but she came back with the argument that because science helps us understand the universe, and that because it is so increasingly intricate, it must've been created by a divine hand. I held my tongue out of respect, but I wanted her to understand the law of the infinite, a basic law that most people can't seem to grasp. In the law of the infinite, given that you do have infinity at your disposal as the universe does, anything that might happen, can, and will happen. Again, we were on the road. **After much reflection, here's my idea of how science and deity work as one. Natural science is what I call the daily operations of the universe, everything is completely perfect down to the details beyond comprehension. Down to the quarks within the neutron, within the carbon atom, within the compound, within the DNA, on, and on, and on. Although we can't accurately measure everything with the instruments scientists use now, there is no doubt that it, in itself, is perfect. The only true perfection we know, that which governs existence itself, is what I consider god. Not necessarily from a Christian, Muslim, or Judaic view, but with the odds stacked against us as a species, life shouldn't happen. Science is god to me, and god is science. Hand in hand, duality in everything. :)** She gave us snacks, but told us she needed to stay in John's apartment to clean and that she couldn't give us a ride, which was fine considering we expected nothing from her. Only but an hour later, we met the single-handedly most amazing person we'd met on the entire trip. She took us the entire way from

Meeker, CO through to Lander, WY; damn near 300 miles in her makeshift home of a minivan, in which she created a bed stand out of plywood that fit only if all the back seats were down, with underbed shelving for all her needs, complete with solar shower and climbing equipment, she was set to go anywhere and do anything. She was our age, just finished her first year of college as an art major at MSU in Bozeman, MT. She had taken a year off to adventure, becoming a river guide, only but a year or two before that, she surfed, paraglided, rock climbed, and even planned with her friend to climb El Capitan, the largest rock face lead-climbing route in Yosemite, and the greatest feat of pure rock climbing in the history of the sport. She was everything that an idol could be, and Jake and I were both smitten with her outlook of positivity, adventurous spirit, and artistic ability, creating a simply amazing woman. We spent the next four to five hours talking with her about love, the thrill of adventure, advances in science and the impact of morality thereof upon society, her rafting experiences, swapped traveling experiences, her experiences with men **(obviously followed by apologies for all the horrible guys out there. I want to reiterate, most things I hear about men make me ashamed to be a man, and although I would never change myself and am content with being a man, I feel shame for the entirety of men's actions. Did you know that almost 80% of violent crimes are committed by men? How neat),** and her die-hard love for high-lining (slack-lining along high crevices/canyons), where she found a calming solemnity of spirituality in Moab, and where she felt as though when she fell from the line at night, she was falling amidst the stars. She was easily the type of woman you could never get bored talking with. With a restless spirit that never settled for the feeling of content that most so easily rely upon in their lives. **Why should we ever accept being content? Isn't there more than just the secure and average? We all deserve what we feel we deserve, within respect to the infrangible rights of others, and to accomplish that goal is to fully embrace the type of person you want to be. No one should ever feel like 'good enough' is truly good enough. Settling is not the same as living.** Just a shining example of the beauty of humanity. Jake's type of woman, hands down. **That's why he sat up front with her while I lay sprawled on her mattress in the back.** We got Jake a new phone at the local Verizon store where Colton informed us on the difference between high and low tax, in a nutshell what's essential to survival and what isn't, respectfully. As well as told us where the city park was, which just so happened was holding the local IPA festival. **I remember another thing he shared with us was the stupidity of the common man. Given how close we were to Yellowstone, people came through this town from all over and one man in particular wanted to get a picture of his kid** with a buffalo. He didn't stop there. He mounted his infant on a wild beast of the plains, who not having known anything of the sort, bucked the kid and impaled the father. Colton said the baby survived, but that he heard the dad didn't make it. As for now, we can only speculate where that kid is…probably learning the ways of the buffalo…like Tarzan. Please do your homework, and for fuck sake, don't mess with any animal larger than a dog, or even most horses for that matter. Venders from all over the Midwest came as well as an impressive live performance that focused solely on the percussion/instrumental side of enjoyment, and the air of family atmosphere and great folks made it a clear choice to retire in this town for the night, having travelled well over a quarter of the remainder home. Essentially, it embodied the classical rhythmic festival music that one could draw the comparison in situation to the rhythms of the gypsies who roamed the 1800's.

Days 21 and 22: June 9 and 10, 2018

We woke up to the party starting back up for the rest of the weekend. We slept on the one patch of grass not taken by drunken tenters next to the baseball field, which in part, reminded me of how I couldn't wait to be back home to be a part of my summer team, mostly did my mind wander onto thinking of home. I needed to do laundry too since I was down to my last clean pair of undies, a clear indicator. So, I dragged Jake to the laundromat, reluctantly he waited outside while I put all of my clothes in a washer machine, threw the shirt I was wearing in there too, met eyes with one of the most attractive and active looking fully grown women I've ever seen while she waited for her clothes by writing in a journal of her own, who then proceeded to buy soap, and later telling me that to do so required a special swipe card, as did every machine. **It has come to my attention that I talk about a lot of women. I want to make it clear that in no way do I reflect on them for selfish gains. Each of these women, Donna included, who I consider to be some of the most beautiful women I've ever met are truly that, and each in their own right. Everyone has their defining characteristics, and being the charmer I was raised to be, I recognize them in every woman I talk to, some more so than others. True confidence in who a woman chooses to be is just as beautiful as any physical attraction one might feel, if not more so as well. It only seems fitting that, when brought a fresh perspective of the nature of things, or face to face with someone who changes the way you think, they deserve as much attention to detail as possible, thus their defining characteristics and resulting beauty. The flavor left forever thereafter on one's mind and heart is true beauty.** She offered to let me use her swipe card, and I transferred enough money for my clothes as well as a little extra for her. Donna was her name, one of the most charismatic and confident women you'll ever have the privilege of talking with, that is, if you can find her. A graduate student who studied social working, now in her mid-thirties, only five years ago quit her job social working, sold her home, and started bicycling across the country, leading tours and seeing it all. This was the woman I sat with for that hour. She came from North Carolina, but was in Lander until she completed her EMT training, then she was to go back, on her bicycle, of course. She had been to 50 countries, spending genuine time experiencing each culture, learned multiple languages and was basically the woman Indiana Jones. Even Jake was awestruck at her social adaptivity, and her ability to make a life where few dare dream to do so. Some of the best words of advice she gave us was after we told her how we thought we 'cheated' by riding a greyhound through the middle plain states, but she rebuked it by saying, "You can't cheat at your own game."

That really stuck with us, **still now I say it pretty often and reflect on her principles.** She even offered to write us recommendations for her bike tour company if we ever wanted to join. I finished my laundry, both us thanking her for her time, but while walking away, all Jake could talk about was the bountiful opportunities to make money from the experiences and connections we make, by building a solid social network. He is very monetarily driven, which is a good skill in having the ambition, but to an extent. It seems at times as if his entire worldly perception is based on what can benefit him. **This mentality will undoubtedly carry him far in his life, but towards the final legs of the trip, this fundamental difference in our approach on the world was what sparked animosity for me.** College really did a number on this kid,, because now the conversation topics he approaches follow suit with a matching tonality that implies that he is better than most, even though I personally believe that he is better than most, he even admits to telling himself everyday that, "I'm just as important as the moon and the stars."

Which I admire in a character, but that's not my reality. I like myself, but daily I remind myself that I'm no better than anyone else, of the realization that I'm really just a piece of shit no matter how many beautiful ideas I might have or appreciate. The degradation of the personal pedestal that so many put themselves on allows me to try to retain a sense of logical, emotionally unbiased reasoning that I value so. I act every day for most of my conscious life, so much so that I don't even know where it begins or ends, or even where the lines blur between who I am and the person I show to the world. Jake and I would be great actors, though he feels he could only play roles that portray his character in real life, I try to gain perspective of my situation to be able to slip into whichever role demanded by my environment. At that point in our conversation, Gary picked us up. A simple man by nature, slow to speak, and very direct with his words. He offered us some pizza, but all I could see was the pizza sauce on his teeth that he as a grown adult couldn't feel. I amuse myself by pointing out flaws in society, or even simply put, in people themselves. **I keep my thoughts to myself unless I feel like that person could inadvertently harm themselves or others psychologically/ physically based on their beliefs or characteristics.** People are so damn amazing though, it's easy to stare at someone for five minutes and find a reason to love them, hate them, and laugh with them. **Give me a day, however, and I can find you a thousand reasons to be mad.** He carried us just under 10 miles to the Native American land reservation. Which spanned roughly the next 45 miles of highway. He was busy hauling horses, but warned us about the crime in that area, only to go along with his ways not but a moment after. A lot of people actually warned us about the dangers there, sharing their fears in the hopes that we share in the same. In that sense, it was a lot like the South, where stereotypes are perpetuated and the people are trained into a mental state of docility through fear. Only a few miles into the reservation, Sky and Lynn picked us up, and their dynamic fascinated me. Sky talked about his first wife multiple times as if he still harbored feelings for her, but almost forced himself to settle for his current wife in the passenger seat. She, who being a short, lovely Native Hawaiian woman, had an equally cute little poodle-like floof ball dog that constantly sat in her lap, wavering only before realizing he didn't need to interact with us. He, his father, son, cousin, uncle, and ex-wife were all attorneys. Sky himself was a public defense attorney making the case that everyone deserves representation, which is fair in theory, however, he told us the story of one of his defendants. This man found an abandoned car, broke into it, tried on some women's clothes he found at a highway rest stop, only to be startled by a real woman, where soon thereafter he slit her throat 'out of panic' then beat her husband within an inch of his life, leaving them for dead in the bathroom. Luckily, someone found them soon after and they survived, however, Sky got him off with only a charge of burglary. The way Sky told the story was as if he himself were put on trial and repeated the same lies he had rehearsed hundreds of times in his head during that case, not only trying to convince US that that particular man was just an unfortunate guy in an unfortunate situation, but that he was trying to convince himself to ease his own conscience. That man, not Sky, is a man who doesn't deserve a second, or even third chance. **He made a conscious decision, and acted upon it. He signed his own destiny at that point. You can't just say oops, you have to deal with the fitting consequences of the penal system. Pun intended.** They dropped us off at a gas station in Dubois, WY, just before the Teton National Forest, just south of Yellowstone, around 4. We ate lunch and as soon as we were getting our stuff to get back on the road, the goofiest motherfucker anyone will ever meet in their lives offered to take us to the southern gate of Yellowstone park. And goofy in the sense of a car wreck mixed with a comedy show, the stories he were telling were so far-fetched, yet explicitly detailed to a fine degree, that we genuinely couldn't tell if the man was crazy or in fact one of the most travelled and educated men we've ever met, and I don't quite know if my words can fully do him justice, though I'll try.

His name was Donny Lewis, and his wiener dogs were Oscar and Yoshenheimer (or some other Latin shit that translated to 'otter-dog'). They had travelled, the three, all around the area a great deal of times in their 'Interstate RV™,' him being quarter-Shoshone, and raised in the area, not only showed us his magnificent collection of arrowheads, with a varied assortment of petrified objects, he pointed at damn near every geographical location and had a story and a woman associated with each. Initially, he pointed to a mountain and claimed it to be named after his great-grandfather who caught Butch Cassidy, eventually becoming the first Marshall this side of the Mississippi. Given Donny was in his 70's, relative to Jake and I, that would be on the same level as our great-great-great-grandfather, which seemed practical. **Sarcasm, at its finest.** Then he went on to tell us about his college career as a pole-vaulter for University of Wyoming, leading into his career seamlessly as a second-string quarterback for the Raiders for a year, making the mature decision to play with the Golden State Warriors basketball team from '87–'89. The only truly convincing parts about him were when he said he came home to take care of his mother, coincidentally who had invested $150,000 into Microsoft in 1984, so they were hella rich. That and his connections with the locals, naming who lived where as we drove past their houses. **Not entirely convincing still, but what made me re-question his validity happened later.** Claiming to have helped build damn near every other house, this is only a small part of his list of accomplishments he happened to share with us:

1. Survived two avalanches.
2. Graduate with a B.S. in Chemical Engineering and B.A. in Archaeology.
3. Arabic interpreter in Guantanamo Bay.
4. His brother invented the scoring system for Fantasy Football, and also helped create the internet.
5. Founded Diablo Solar Services, a solar company in California that started with nine employees, growing to 400, being the largest solar company in the world.
6. Helped build Tom Cruise's house in Yellowstone, even taught Tom Cruise how to whistle
7. Worked in Antarctica, but was fired for befriending and interacting with a penguin. Something that is generally looked on with distaste by researchers, who don't want anything to interrupt/affect the animals they're observing.

All the while, still pointing to other places along the way where he's worked as well. Being fired from a ranch for flirting with the several women who came there, in a straight forward and prudish fashion. Then we stopped at the Red Fox Saloon, so he could buy a couple beers to recover from his hangover as a result of the night before, when he blacked out, fell, hit his head hard, breaking an arrowhead display case, and leaving behind a scab that covered the good majority of his crown. For that, I do feel bad for him, in a way of sharing pain and seeing myself in his. The man had no boundaries, and as Jake and I watched the Triple Crown horse race making history with some series of consecutive wins that hadn't happened in the last 40 years, he hit on every woman in that bar. He was devoid of conscientiousness, which was an admirable characteristic. Very childish in nature too, however, constantly making sexual jokes or innuendos, even to alleviate his pain talking about how his friend's son died. He said, "It was hard on all of them…" trailing off and falling silent, pausing for a moment, then laughing at how he just said, '…hard on…' before the emotions began to rush back in. Hearing 'that's what she said,' in the conversation every five minutes had become our normal.

Here were his favorite pick up lines:

1. Are you from Tennessee? Because you're the only 10 I see.
   He said this at least five times during our time together.
2. Are those galactic pants? Because your ass is out of this world.
   And my personal favorite, which is actually pretty good
3. Are your parents thieves? Because they stole all the stars out of the sky and put them in your eyes.

As he dropped us off on the border, he left us with the funniest quote and justification for not wanting kids by describing a woman with three kids by saying, "God damn, it's a vagina, not a damn clown car!"

And that was that. We shook hands with this man who had opened up his reality to us, and for that I respect him. We snuck through the woods around the ticket station, because why the hell would we pay a $50 entrance fee? Then walked the next four miles to a pasture overlooking the Teton Mountain Range. It was extraordinary. We knew from Google maps that the next stop was eight miles away, so we said fuck it and set up our tent off to the side, with a bear hang decently far away. Now, I am writing these two days at the same time because I don't think that the night can accurately distinguish the two, but rather the second is a run-on sentence of the first. We were thankful to have set the tent up when we did, because as soon as we did, no joke, hundreds of mosquitos were on the screen, waiting for their chance at us. Thank the heavens for their own self-interests, because if they had overcome their own selfishness, they could have easily teamed up together to pull apart the mesh barrier with their needle noses. The bear hang was tougher than I thought it would be, only initially, but it took forever to finally throw the rope over a branch high enough to be out of range of any bear. As last, we felt comfortable sleeping on the top of that grassy pasture, looking over the scene cascaded by moonlight, as if it were more beautiful and perfect and serene than anything Jake or I could have imagined. Then the park rangers came. At 9:30, Jake woke up to a car's light bar shining in our tent from the close by parking lot, they woke me up when I heard their boots march over, made an effort to check our licenses and told us we couldn't camp there, however, there was a campground just five miles up the road. They, being of either high prestige or rigid schedule, couldn't expend the 10 minutes to drive us there, so we walked until midnight through grizzly infested territory until we walked across a lodging development, fit with gas station and just enough people, which we thought would most likely prevent bears from coming too close, so we unrolled our sleeping bags, no tent, and fell asleep under the stars on the concrete outside the gas station. Three hours in, only to be woken up again by some asshole shining a flashlight on us from inside his car, saying this was private property and we couldn't be there for liability reasons, ultimately coming to his point, we had to leave. The fucking guy watched us pack up through that magnificently unwavering beam. What makes me the most mad was that he didn't even operate the gas station, he just made sure that we left his precious gas station, driving back to his villa where the only thing he might be able to sympathize with would be with the person looking back from the bathroom portrait he would paint himself. It's seriously three in the morning, we're dead asleep, it's not as if hobos are a common problem in Yellowstone, or a problem in any respect. The worst part was that he didn't even ask if we were okay, or bother to know our story, but like the City Market manager, only assumed the worst and couldn't possibly think outside of his own prejudices. So, we slept in the bushes just outside of the lodging until sunrise. **There was an entrance sign, lit up by many spotlights. Humans are some of the most oblivious animals in the world, with practical reasoning so our ancestors could focus on motion and see large predators, but to the same extent blinds us to everything around the one thing we observe. If you look at a lit sign in the middle of the night, I guarantee you that you don't remember or even see the darkness around and behind the sign, or under it. That's where we slept, behind and under the sign at the front of the park, because everyone will be looking at the sign, never thinking to look for the obvious.** We went back to the gas station and made new cardboard signs that said we were from

7B county of Idaho, now that we were seeing a lot more 9B plates, the county district just to the north of our hometown. We saw our county neighbors pass us by, even people from 7B, who we easily could've known or who knew someone we would, not a single one stopped to show any love. We made it to the next gas station rest stop, **probably what google maps said was initially eight miles away,** around nine where we charged our phones and I flash read both treasure books for more clues. Just as we were about to set off again at 11:30, both of us were needing time to calm down, relax, and just let our bodies rest, another incredibly beautiful woman noticed Jake and I's signs. Pam was her name, presumably the mistress **(we later discovered fiancé)** of the man who she convinced to let us ride with; Bob. They had an interesting dynamic also, they loved spending time with each other, Bob almost seemed to be willing to do anything for her and even joked about marriage one day, which Pam seemed to laugh off, changing conversations. They shared with us their experiences, mostly in a Bob→Jake, Pam→I fashion, where Jake and Bob talked about entrepreneurial type opportunities and Pam and I talked about places we've gone, swapped pictures, and exchanged Instagram info. They gave us a bag of Mississippi blueberries, Bob's company 'Half Shell Seafood Diner' being primarily based out of there, and some cooked fish they caught the day previous before they dropped us off at Yellowstone Lake. Jake was going off about money, and I finally let my tongue slip saying how I think money stands for everything that's wrong with and what I hate most about society; **a necessary evil as I've come to believe.** He countered with his ideal of how money stands for everything great and essential to society. We didn't say much for the next hour, each touring the lakeside hot-spring pools at our own leisure. After that hour of silent introspection, and staining my pants with the grease from those delicious ass fish **(which after a delicate procedure of grunts and body motions we had devised as a second language, Jake communicated that it didn't smell appetizing),** we started walking again, only to make it half a mile further before Wayne and Marilyn picked us up and took us all the way through Yellowstone, **stopping briefly for a restroom break at Old Faithful, where I branched off on my own if but for a moment to remember watching the geyser through the same window with my parents only 14 years ago,** all the way to the city of West Yellowstone, MT, just outside the north entrance of the park. Marilyn was a traditional woman in the sense that she valued a steady job with security for family life, Wayne having worked as a logger for Georgia-Pacific for his livelihood. A fisherman and naturalist, he pointed out all the great fishing holes and forest deterioration signs as well as animal rubbings on the bark of various trees.

Marks that Jake and I know well from the wilderness oasis we call home, but we still acted as though this was all groundbreaking information, well Jake did at least. Honestly, I zoned out for that hour in the car with them, and all I could think about was the treasure, the irony being how much I detest money, but the realization that it is essential to having a decent life in the world we've all helped build. **I do admit to having selfish thoughts of what I want to do with my life and knowing that I need money to support my future endeavors. In a perfect world, money wouldn't exist, there would be no restrictions, no expectations, save for that of morality and common decency, no legal pressures on those who make our decisions on a federal level one way or another. No obsession with every generation of this nation's youth to have the perfect, unachievable life, only attained through money, and a life of vain blissful ignorance towards how seriously messed up the world is outside of their own selfish bubble.** I also thought of the irony of the lodging facility we slept in front of, where people pay to escape from society and find solace in nature, only to go to a building with TV, electricity, and internet, surrounded by 40 other little houses doing the same things, creating what they were escaping from. The smugness in his tonality too made it hard for me to remain courteous, so it was good to let the mind wander for a while. What I do remember from Wayne were the emotions that came to his face as he stood outside with us, dropping us off and taking a step aside to discuss the importance of an adventurous spirit. He told us how much he respected us, and if anyone ever tells us we can't do anything, to tell them, "Eat a bucket of shit."

After he said that, he paused for a moment to collect himself. Within his collection, Jake and I both clearly saw the eyes of a man who lived with regret. Regret in not taking hold of his life sooner. Regret in not achieving everything he set out to as a child. It's the little things we can't redo, but wish we could. He wished he was young enough to come with us, or courageous enough to dream big. He wiped the tear and walked back to his wife, with whom I hope he genuinely finds peace with in his relaxation age. There's always something to learn from every situation life presents. In this instance, learn to live without regret. I do remember now clearly that they, mostly Marilyn, were slightly disappointed in their 35-year-old son and his girlfriend for travelling the world instead of finding a steady job. Then Jake and I ate some damn good pizza at the same restaurant that had five pictures of Blake Shelton on their walls from the one time he came, played a hunter arcade game to enjoy the simple things we missed, and threw up our hammocks in the park across the street. Only about 20 miles south of the region of forest we believe the treasure lies. Tomorrow will be a good day.

Day 23: June 11, 2018

We walked a ⅔ marathon today. We didn't talk about much, mostly due to the freezing cold and rain we were walking through all day. We had only really talked when we figured we absolutely had to, as in telling the passing cars that we were gucci, and that we actually in fact enjoyed walking through the freezing cold. We spent about an hour in the morning looking at a map of the forest we were about to walk into and be swallowed by. This was where Forrest Fenn spent his early-childhood, and we pieced together the route he would've taken with his friend relative to the poem/clues he wrote about in his book *'The Thrill of the Chase'* in which he gives vague directions to the location of roughly $2,000,000 in gold that he hid in 2010. I traced a radius circle around the valley he started in, only to find five exit points possible. Then checked them relative to the tallest peaks, which he claimed he and his friend to be just north of, narrowing it down to two. Then because I knew Forrest was a pilot, and saw that only one exit point was close to an airport that he would've used. Then we followed his set of clues based on that location, and it started matching.

1. "Below the home of Brown"
2. "No place for the meek"
3. "There will be no paddle up your creek"
4. "If you've been wise and found the blaze, look quickly down…"

So, with that and both of us feeling relatively confident, I gave Jake a location 50 miles away and let him lead the way. **Not to mention, we had both almost died the night before in our hammocks. The 30 mile an hour winds running above and below us, mixed with the cooler nighttime climate of the Rocky Mountains made for a cold night through down to any man's core.** Little did we know that the five-minute difference between routes on Google maps would more than likely cost us the rest of the day. And so, we walked. And walked. Brady gave us a ride for three miles, being interested in physics and the world around him, he gave us food for thought which I'll attempt to explain now.

Question 1:   Can we oxygenate space? As in the space around our planet.
My answer: No, we wouldn't be able to oxygenate space, mostly because there simply aren't enough atoms to fill the infinity of space.
Question 2:   Could you rip a hydrogen atom in a vacuum? I.e. space.
My answer: Theoretically, you could if you had the right equipment, however, the force it would take to tear apart the strong force of the nucleus would be the likes of which would cause a gravitational pull in the form of a black hole, potentially. That's how we make atomic bombs without a vacuum.

Then we kept walking down the loneliest highway in the whole pan of South-Eastern Idaho, making our way to the part of the forest, 50 miles away, where the treasure waited. Our only saving grace was an RV-Marina-Store where we ate lunch at 2:30, looked at some very interesting zombie-prep weapons, even a Nazi memorabilia dagger, fit with engraved swastika, eagle and all. Then we set back off down the road until 5, got a four-mile ride from Jason, who told us the folly of our ways in taking this highway instead of the other two busy ones, then dropped us at the intersection of the next state highway he claimed to be supposedly busier, though just as lonesome, if not more so merciless. We set up camp along the Madison River, only 24 miles south of the forest now, mountains, and our treasure. Although it was arduous, and our shortest mileage day in the West, our spirits were still fired up, and we got an amazing night of sleep, falling asleep to the idea that we might find treasure the next day, only if the god of roads decided to shine mercifully on us.

Day 24: June 12, 2018

The gods were not merciful. We walked 21 of the remaining 25 miles today. We stopped at another RV-store-campground early in the day, which are actually very popular in Montana and every person you speak with either leads fly-fishing trips or skis, and we finally used the restroom after holding it in for the last three miles. We would've been stupid not to buy more cheap snacks, which I'm very glad we did because there was no other place to do so for the rest of the trip into the mountains. We were about 20 miles away, and walked for the next 15. That was enough time to do a sprint with our bags for just shy of a quarter mile, uphill, more like 5/32 of a mile, but still, and for Jake and I to help evaluate each other's character into three steps: (especially towards women)

1. There are two ways to raise yourself mentally, raising others around you through support, or cutting down others to make yourself look better. I start any relationship by cutting myself down to help raise other people, so they can boost themselves morally. **People always need a reference, and as long as they're doing better than at least one person, they'll feel better.**
2. I 'manipulate' people with the intent of helping them create the best version of themselves. In the case they come to me with a problem, and I try to help them overcome it. Everyone has the potential for greatness.
3. My confidence facade creates the illusion that I'm higher, or better than most, which some might see as a goal to strive towards. **Become the role model you wish you had.**

Jake's character analyzed into three steps:

1. He tries to show as much of his personality as he can as fast as possible to try and connect with people.
2. He evaluates if they're worth his time.

3. He is always looking for something better, which could incentivize others to do the same and become the person he knows they can be, but never gives himself time to truly slow down and find all potential with what's around him. He pretends not to care in order to make a person want to better themselves, hopefully left with a goal in their minds that one day he will care. **But he does care, it's just super-duper secret.**

I find solace in not caring. I like both ends of the caring spectrum, caring blindly and wholeheartedly as well as caring not. I like to help and invest time with people given Jake's 3-step evaluation of me, only if I can see their potential to recognize my actions, maybe not exactly appreciate them, but have the capacity to replicate what I do for others in the future. **In short, I want to encourage others to help others, but that starts with helping themselves.** For women especially, I want to raise their morale and esteem as much as I can so that in the case a relationship doesn't work out, they hopefully leave stronger, more empowered women. I do so by cutting myself down initially, remaining humble and making my accomplishments out to be not worth much. This makes anyone feel higher in comparison, all the while I masquerade as someone better than I know I am. I know I'm some half-assedly created goon, and find solace in reminding myself of that truth. But if my confidence falters, if I let my guard down, or start to care too much, the whole effort collapses on itself. **If this part wasn't written in my journal, I wouldn't be writing it here. As I am honest, I feel I must write as similar to my personal journal as possible, I never had the intention of others reading it, but I was not the one on that trip, the man who journaled was.** We talked with a camper at the campground that morning, who ended up driving by us on the road and gave us a ride to the next store before the mountains, where Jake bought a badass Damascus steel knife folded 512 times. **My grandpa instilled a sense of knife enthusiasm and appreciation in me at a young age. He and I just forged our first knife from a railroad spike last month.** They were all so beautiful, all made individually and all with their own unique wooden finish. It took a lot of effort not to buy one. We were four miles from the base of the mountains, and I bought another pair of sweatpants, having bought a sweater and thinner sweatpants/pajamas immediately after a miserably cold night in West Yellowstone. We hiked in, now only about a mile from the base of No Man's Peak, set up camp along the closest creek by bringing huge rocks scattered across the small field into a circle where a fire was made previously, now creating a fire pit, spending the next hour collecting firewood. It didn't catch on the first match, but with a little adjustment, we had a decent fire roaring in no time. About half of the wood was wet from the late spring rains, filling the canyon with smoke, making our presence known to all creatures who call this place home, as well as doing our part to clean up the forest of its fallen comrades. We threw up the bear hang again, Jake made a farewell video, I had just messaged all of my loved ones earlier because there would be no cell reception later, and we romanticized finding the treasure tomorrow. Never had it been closer or more real. **In retrospect, I feel like we had to prove ourselves to the universe that we were worthy of finding this treasure by walking the last 50 miles to it.**

Day 25: June 13, 2018

It is very clear that Jake and I have different goals for this trip, and this miscommunication has caused a rift between us. It was my understanding that the main objective, or at least what we've been telling everybody thus far, was to find Forrest Fenn's treasure hidden in the Rocky Mountains. It was a bitch and a half of a climb today with our packs, but Jake made it look easy. He's the type of man who doesn't look back when he hikes, which in itself makes you want to push yourself to keep up with him, while also leaving a sour taste in your mouth. He disclosed to me after refusing to eat the dinner I cooked us that he honestly couldn't care about the treasure, he just wanted to see the lakes at the top of the mountain and hike. It frustrates me how he walks so confidently and fast, with his head focused on the trail, only thinking about the destination and the future, coming back to reality only for survival against predators that could have easily been watching us. I had been going slow and steady all day, mostly because I was pretty heavy as it was, but also because I was making an effort to remain calm and focused, trying to look for the treasure wherever it might be. I've been very gung-ho for the last couple of days on it, and to Jake's case I had let the climb affect my attitude more than I'd like for it to have, but the idea that I will never have to be dependent on anyone with this treasure, creating stable investments, and being able to run away with the woman I can't help but fall for to some town in Spain that she always romanticized, for that idea was too alluring of a call. I've been looking down damn near every step, and it truly pissed me off to see Jake with his head down, not even trying at the already minimal chances we had to find it. I swear, he could've walked right over it and never been the wiser, but it isn't good to dwell on such things. The wind last night kept playing tricks on our minds as it bellowed through the canyon and our small field we set up home on. At first, we were on edge by how the formation of sound recreated the beating of native drums and war cries, but the more we thought about it, the more the sound had transformed into the laughter and cheers of whom we could only assume those fallen, as if their spirits live on through the mountains, living in an eternal state of merriment with their loved ones. That thought was more comforting than the potential one of us camping on a native burial site, it being a quite beautiful pasture on the easy bank of the slower section of river, with various pockets of large rocks, though not in a seemingly noticeable pattern. **Had we just removed the grave markers of the dead? Who knows?** Jake almost left at dinner tonight after I made a joking comment. He was talking about how hungry he was, and being the ass I am, I said without hesitancy, "Guess you should've prepared better…"

In all fairness, the conveyance was pretty indignant and smug, leaving an even shorter fuse than the one that had been burning away the whole trip inside Jake. I had five meals of freeze-dried food left, and our new camp for the night was halfway up No Man's Peak, along No Man's Creek, where we had a never-ending supply of freshwater rolling by us. I was also carrying a couple snacks, and he was almost out of food. **I had told him there was a shop right before we got into the mountains, and there was. However, they didn't sell food, it was only for their fly-fishing tours. Jake had put all his eggs into that basket, which I had accidentally led him to believe sold food. On Google, they were labeled with a food mark, but I never put all my faith in anything, so I prepared for the situation where they didn't sell any.** Even if we don't find the treasure, I highly doubt we'll be on the road much longer. I'm close to calling a family member and asking for a ride back home, we were only four hours from home by car now, it wasn't an unrealistic possibility. I hate being dependent on anyone, with every fiber in my body. I don't like the idea of quitting either. I genuinely hope one of the following two things happen:

1. We find the treasure. Get ride home.
2. One of us, preferably me, gets attacked by a wild animal, though not nearly fatal, just enough to limp out of the woods with a sliver of consciousness so we can have an excuse not to be looked down upon for leaving.

I miss my family. And my brother. And people that actually kinda give a shit about me, and wall outlets, and food, but mostly the ability to go somewhere by hopping in my Jeep and just going. **That's not to say Jake doesn't care about me, we do care for each other like family, but it's different when he's the only one who cares.** Jake and I can agree on what we miss. I hate repeating myself too, which is ironic because I usually talk on the quieter decibel levels that most aren't trained to hear for, but this entire trip feels like a broken record of the same bullshit questions that everyone asks about 'you' only so they can feel good about themselves for 'caring' where are you from, where are you going, what's been the best part, are you in school, why are you doing this. It'd be different telling people that actually care, but we both know damn well that deep down, at the end of the day, they couldn't give a rat's ass about us, and in one month, if not sooner, they'll forget us entirely, save for the fleeting memory here or there throughout the remainder of their lives with, which no precedence or truer meaning holds any value over them.

Day 26: June 14, 2018

Jake and I are descendants of two-character lineages as old as time, the dreamers and the realists. Although they butt heads at times, they need each other to survive, the dreamers need the realists to make sure they don't fly too close to the sun as Icarus's son did, and the realists need dreamers to help them get off the ground in the first place. I thought the day was off to a good start, Jake seemed to sympathize more with my romanticization of the treasure, and for the first couple hours, we were good, mostly until we got to the top, where Jake allowed his hunger, pain, and, by his account, sleep deprivation affect his outlook on the world. When your animalistic nature breaks down the character you show to the world, it shows who you really are. Over the last few years, I've always thought that between Jake and I, he was the realist and I the dreamer, but when we summited No Man's Lake, I was quick to see that he had all but given up looking for the treasure, even given up the hope that we had a possibility of finding it while I was searching high and low. As we were starting to go down, my attitude already at a low, he looks at me and says, though not exactly, 'I don't want you to resent me for not looking for the treasure,' trying to hand me his bear spray, as if he was convinced that what I wanted was to go off on my own further up the peak to keep looking for the treasure. **In his defense, he was only looking out for me, and after review now, we believe the treasure to be somewhere different, but it's the principle of the matter.** Friendship is more important than gold, and even though we both knew there was hardly a chance of finding it as it was, he doubted my capabilities. That pissed me off to no extent. It disheartens me when someone doubts themself, but never doubt me and what I can do. I went down so fast you would've thought I lived there, knowing the terrain as my own. It's funny how so much can change in three hours. Right then, it started to rain. Sometimes I feel as though the weather has a tendency to mimic my current mentality, even though I know that's scientifically impossible, but still kinda neat coincidence. As I grew with intensity of thought, the rain equally as fast transitioned to heavy wet hail, and the downpour started. I had calmed myself going into base camp nearly half an hour later of running/working my way down the mountain on my own, the rain at that point had come back and slowed to a light shower. I ate my can of soup, and granola bar that I saved for Jake, although he was too proud to accept it and would never eat it even while his hunger ate away at him. At camp, we both agreed it best to leave the mountains through our grunt language. We walked the next eight miles, four hours of which were uphill, 3.5 hours of those four were during the most intense thunderstorm/hailstorm that Montana must've not seen in a good while. Not a word was spoken for those miles, though eventually through the sharing of pain, we were on grunting terms, and when we finally escaped that wretched mountain valley, through which we had to go over another mountain just to go through the pass, my shin splints came back, but only subtly in my left leg. I don't complain about pain, and frankly I look down on complaining as a sign of weakness, but throughout my entire life, shin splints have been my greatest enemy by far. For those of you who don't know what they are, they are a result of certain calf muscles not being exercised as much as the others, to the point where your bone literally starts to feel fractured, or splintered. We took a rest at the creek campground, luckily there was a retired Alaskan Hunting Guide who offered us provisions and cold beer to last us through to the next town, **though I could've done with a lukewarm beer at that point to warm up.** I was grateful and ate some of the chocolate he offered us and a minimal sandwich, one slice of bologna, one of cheese. Jake on the other hand had all but forgotten courtesy after his hunger got the best of him, eating three sandwiches and half of the pack of bologna, inhaling cheese as he went. I reminded him to only eat enough to get to town where a mom and pop diner/campground waited for us, but he couldn't care less while I drank my beer in contempt. In our conversation with Bill, I could see how Jake throws around the word 'we' too loosely, and emphasizes points on our story that don't really need emphasis.

"We were hungry."

I wasn't, I was prepared, and even offered my food to him.

"We didn't find the treasure."

Well, first you have to attempt it.

"We were tired."

It takes a lot of daily practice, but I make an effort not to let my sleep or physical exertion affect my outlook on the world. And when he tells our story, almost always one of the first things he says is how we 'cheated' by going greyhound, after going off about how great Donna was and her words of wisdom, even continuing to do so still, which makes our 1500 miles feat (excluding the greyhound) seem worthless off the bat. I'm sick of this, I miss my family. After talking with my parents, who had already committed themselves to our summer baseball team, and who were hosting a wood bat tournament this weekend, I decided to stay another few days. I talked with Jake and he says he already feels at home, though 400 miles away from our actual home in a merciless domain, the only connection to which he has is his college 200 miles away. This kid is so damn oblivious that I genuinely feel he wouldn't have left the mountains unless he had someone to say, 'Hey, this is the trail to get out of here,' while he had already started down the trail that only would've taken him further into the torrent. Or someone to let him know which roads we had decided to take so he doesn't end up talking with the person driving for so long that he ends up further from home than he started. **Our ride with Brady on Day 23 was close to becoming that situation.** His blinders are always on, and I would feel like shit for abandoning my brother for dead. He didn't even prepare for a mountain trip, after shopping at a convenience store (Marina/Store) the day before, the thought had never crossed his mind that the next place might not have food. Plus, what would I say to his mother.

Day 27: June 15, 2018

Fuck it, we've already gone 90% of the trip, what's another week and a half? I like to live my life without regret and plan my future accordingly. There's going to be more Father's Days (hopefully), but if I leave now, I'll always live in regret for not finishing something after being so close to the end. We decided to stay tonight again in Cameron, MT, at the Blue Moon Saloon, a smart decision given that we both needed to reground ourselves, me especially to let my shin splints rest. But all this time, staying still has made me reflect on the main thing I regret most in this point in my life. Hindsight is always 20/20, isn't it a bitch and a half? All the time I spent with EMSRP up until the final night when she got all dressed up to an extent that no woman ever has for me, just to spend time together doing homework, going for a drive, dance in the dark outside of a bear enclosure (which I wish I would've slowed down and enjoyed every second of, but couldn't because I was too happy to perceive the moment as more than being present in it), to the movie theatre where I held her hand for the first time, kissed it, and didn't let go for the rest of the night, to me dropping her off at her car, giving her the fullest hug I could muster, not asking her to stay with me until we couldn't stand any longer. I should have kissed her. I haven't seen her since. **On fun coincidence, nonetheless, she applied for the same job I have.** I should've proved to her that what I felt was genuine, and that she was worth enough to ignite a passion to the point where she would've completely forgotten about all restraints society had placed on either of us, to feel free beyond bound and beyond that which the human mind can translate into words. I could've done more for her, and I love the characteristics she brought out in me. I valued the thoughtful and calculative way she approached the world with beauty and grace the likes of which I've seldom been privileged to. I miss her and her smile. She's blocked me on every social media since, which speaks volumes for our generation, and essentially let me know I should never talk to her again. **The background of which is I met her while donating plasma, and was awestruck instantaneously. I asked for her number, she fumbled over finding the right words, later telling me she was just as, if not more so, nervous. She forgot her own last name for a second while I got her phone number, and** for 2–3 weeks straight we talked. **On our first date to a coffee shop, I came in about five minutes late, walked past her to the coffee counter because I didn't recognize her all dolled up, making an effort to look nice. I spilled coffee all over myself when I sat down, and there began our conversation that closed that shop and the next bagel shop down the street. I was more comfortable with her than anyone I've talked to since becoming an adult. She then told me that she had a boyfriend, but that she wanted to keep spending time together. I didn't care what the conditions were, I could not get enough of this woman. Studying then the movie theatre was our third date, and I can still remember the smile on her face walking up the stairs to where we were doing our homework. Never had I felt as confident in being somewhere as I did then. Then we started getting too close, and like I always seem to do, I was 'too nice.' My relationships can be surmised into two groups, the ones I don't put enough effort into that I don't take seriously, and the ones I take too seriously that I put in too much effort for normal people. I take my time in the relationships I value and want to build, but almost painstakingly so, to the point where ⅔ of the women I've had true feelings for won't even talk with me. It's fucking great.** I updated her nearly every other day on how the trip was going. **I know it sounds pathetic, but at that point, what else was there to lose.** She had just started ignoring me a month before this trip started, and the only way I deal with people who ignore me is to pretend like they aren't ignoring me, sharing with her my personal collection of scenic pictures from this trip that I hadn't planned to share with anyone else, but whether she reads them or even cares is beyond me. All I want in this instance of sheer pain and misery is to run away with her and forget the world. I think I might have been in love years ago, but it was mostly infatuation with the potential woman she could be. **She is a fantastic woman in all regards and we are still on good terms.** I'm head over heels for the woman EMSRP is, and continues to be. I bought her and my mother a pair of earrings, hopefully I'll be able to finish the trip to bring them to both. When there's nothing else to reflect on but one's self-pity, I can't close my eyes without seeing her in her floral maroon dress that cut off just above her knees, clutching her hot chocolate with both hands, and her smiling that damn smile. As I sit here, smoking a cigar, listening to Chris Stapleton, I feel like a man who wants to cry, but my life has brought me to becoming a man who cannot seem to find the tears any longer. Maybe the constant drip from my nose for the last five years compensates for my inability to do so. We start again tomorrow, hopefully my mind will be preoccupied with the pain again.

Day 28: June 16, 2018

I don't usually throw up, and when I do it's only if I make myself. By three in the morning, I woke up to my body telling me it was time. Still not sure if it was from fully inhaling the two cigars, which was great in the moment, come back to bite me, the homesickness, the self-anger, or the endless shit that I've been shoveling into my mouth to try to fill the hole of heartbreak and to feed my sugar addiction. Accidentally bumped Jake coming back into the tent. It didn't stop raining all day, and at seven, when we were soaking wet inside our own tent, we decided to dry our stuff out in the bathroom of the Saloon until noon, or at least until we were tired of waiting long enough to say fuck it and start walking. Jake wanted to go down to a hotsprings further south in Idaho. It being Father's Day tomorrow, I expressed my concern for wanting to be home sooner rather than later, but mostly held my tongue for the day. We got four rides in total, one from Jeremiah, from Wisconsin, who just wanted to see the west coast and lived out of the back of his truck that he also turned into a living space, and where I spent 20 minutes in silence. We started walking out of the town he dropped us at, simple and sweet. We made it about six miles out of town then another man and his dog let us jump in the bed of his truck, taking us the next 15 miles uphill, which we were damn grateful for. He was on his way fishing, **something we figured since there was minimal room in the bed of the truck next to his inflatable raft. Foreshadowing?** Though short in stature, but fully man enough, complete with matching manly beard and all. Immediately after jumping out of that bed, the next truck to pass by let us ride in the bed of his truck too. A construction truck, flat bed with handmade wire walls surrounding the bed, that had just freshly been used to transport rocks and debris (as observable from the coarse grey powder/rubble still left over) allowed us to see and appreciate the surroundings we passed by. **I think that not enough people waver their gaze from where they're going to what they see around them. Only when that sense is stripped from you, and you allow the wind to carry your footsteps do you see around you.** We passed through Virginia City, like a piece cut out of time from the Old West, elegant in the retention of its culture, it had even been a trading post since before Lewis and Clark established their legacy. From there, it was another six miles before a guy and his friend took us five more. **By this time, I was so into my music that I don't know how Jake didn't run across the road and break my phone. I had progressed well past country music by this time in the day, and was working my way through the best funk from the '70s out there. Strutting down the road, anything to take my mind off of those damn shin splints. We walked the next five miles to another small town, where the state highways seemed to all branch out from. We resupplied inside the local store, and it being a small town of 10 buildings on main street, got our fair share of sideways looks. When we finally got into town, I needed a break, just to stop and rest, and we made do for a little while behind a gas station, though they didn't have everything we were looking for, like fruit. Making our way to the local store, we got everything we needed, even refilled our waters for free. Only once the entire trip have we ever paid for water refills from gas stations, more often than not, they are super gracious and understanding, and let you fill up your bottles for free. Finally stopping a third time on our way out of town next to a slow, casual river with a park situated right on the banks for just an occasion. We charged our essentials, spent another 45 minutes stretching and relaxing, then hit the final highway.** We walked to mile marker 23 on the way to Dillon, MT. The numbers had only gotten smaller and smaller and we knew that without a doubt, we were going to make it to Dillon the next day. A cop came in at around 11:30 to check on us after receiving several, not just one or two, but several calls about seeing a tent on the side of the road. Apparently, it's 'illegal' to hitchhike in the entirety of Montana, of all places. Jake and I joked for the next half hour, the cop letting us off with a warning, about how we probably have another cop coming right now to kick us off the road, making up what we thought the calls sounded like.

'Eeedgar! Eeedgar! There's drifters outside!'

'It's a drifter farm, Edith. There's gunna be drifters outside!'

Day 29: June 17, 2018

Happy Father's Day. Today is my last day on the road. I called my grandpa, who picked up on the first ring, getting in his truck almost immediately. **This was a man whose kids haven't talked to him in 10+ years for reasons of their own pride. I'm the only member of the family that still reaches out to him with love, and it being Father's Day, him being the father of my sperm donor seemed fitting. I hope he appreciated our time as much as I did.** I'm very fortunate to have the family that I do, and I can confidently say that I'll do my best not to take them for granted. With no other goals to truly accomplish, and so many other things I'd rather do with my time, I feel a lack of control of my situation, which only makes me more frustrated and upset. I like to be able to do things, and go places, and live life the way I choose to live it. It seems whenever one of us is in a pissy mood, the other feels opposite to the extreme. This only creates tension. Jake is the perfect embodiment of the characteristics great about our generation, but perhaps to a fault. He is very prideful, self-driven, self-motivated, enthusiastic, and very capable of damn near anything. His self-driven approach, however, makes his rationale based mostly about how he can control his environment to his advantage, surrounding himself with people who are advantageous to him as well. We got in our largest fight for a good 20 minutes as I expressed my growing desire to be home, however, I felt it my responsibility to make sure he got home as well, **though I wasn't as articulate, the point got across.** He presented me with an ultimatum, either I leave, he makes me leave, or I enjoy life with my brother in arms. **I tried to take his approach for a short while, but I was in no shape physically or mentally to appreciate much.** He leisurely walked through the muggy overcast morning, taking it all in, which I can appreciate in a person, but my judgment was too inwardly focused to appreciate the beauty. I almost felt it my job to take care of him for the next couple days, but he made it abundantly clear that he didn't mind leaving me behind so long as he was doing what made him happy. **And for that, I respect him. For the hour and a half or so before our first ride, I watched as Jake felt he needed to stretch more and run, so off he went, up the road and around the bend. Gone from my sight on mile marker 10. Around mile marker eight, I was picked up by**

300

an older man on his way to the bar, or work, I honestly couldn't tell you because I paid minimal attention. I was done. I let him know that my friend was just up the road, and after picking Jake up, we sat mostly in silence, the three of us. Jake was trying to be energetic, trying to get me fired up about keeping on keeping on. I was not reciprocating anyone. When I told him I was leaving, his face was mixed with both sadness and relief. Sadness in the fact that I couldn't overcome my own personal inhibitions to continue with him, but relief in the extent that he could enjoy the trip fully, how he wanted, unbound by anything, a truly free spirit.

I'm disappointed in myself. In part because I wasn't able to mentally or physically finish the trip, as well as in my inability to escape my own selfishness. But mostly because I was inconsiderate to the wellbeing of the only person that had my back from the beginning. There are many things throughout this trip that I regret about my behavior towards this man who has become my brother, however, being a college student I have less and less time to spend with my parents. Another realization I have come to on this journey is that there is nothing I will regret more than not spending time with the people I love while I have the capability to do so. **At the end of the day, it boils down to what will you regret most. There's boundless possibilities in this world, and you can kick yourself for not doing them all, but at some point we have to make a choice. Will I be selfish with my time, or will I spend one of my few remaining summers with those I love?** I can't wait to be home, hug my parents, learn violin **(to which my hand cradles the strings perfectly)**, read more, and be the man I've finally decided I want to be.

Maybe I wasn't Worthy of the treasure, but I'll be damned if I'm not Worthy of becoming the man I want to be.

www.ingramcontent.com/pod-product-compliance
Lightning Source LLC
Chambersburg PA
CBHW061503180526
45171CB00001B/20